Molecular Machinery

For Alison

Molecular Machinery

THE PRINCIPLES AND POWERS OF CHEMISTRY

Andrew Scott

Basil Blackwell

Copyright ©Andrew Scott 1989

First published 1989

Basil Blackwell Ltd
108 Cowley Road, Oxford OX4 1JF, UK

Basil Blackwell Inc.
3 Cambridge Center
Cambridge, Massachusetts 02142, USA

British Library Cataloguing in Publication Data

A CIP catalogue record for this book is available from the British Library

Library of Congress Cataloging in Publication Data
Scott, Andrew, 1955–
    Molecular machinery: the principles and powers of chemistry/
Andrew Scott.
      p. cm.
    Bibliography: p.
    Includes index.
    ISBN 0–631–16441–3
    1. Chemistry——Popular works.   I. Title.
QD37.S38   1989
540——dc20   89–32200
                                                                CIP

Typeset in 10 on 12 point Palatino
by Dobbie Typesetting Limited, Plymouth, Devon
Printed in Great Britain by Camelot Press Ltd, Southampton

Contents

Preface

This book has a simple aim: to offer an insight into the world of chemicals and chemistry to anyone who wishes it. If you are unsure of why you should want to receive such an insight, I would encourage you to read the Prelude on pages 1 to 4. If you are unsure of your ability to understand and appreciate the insight, I would encourage you to have no such fears—I have tried to assume nothing about you as I have written this book, other than your desire to learn about what chemicals are and how and why they do the things they do.

The view of chemistry offered by this book is of its fundamental principles and basic powers. I do not attempt to provide a comprehensive summary of the chemical world, merely to reveal the foundations upon which all the complexity of that world is built, and some of the great powers which rely on these foundations.

Chapters 1 to 7 reveal the simple principles that underlie all chemistry, while chapters 8 to 19 explore the powers of these principles as applied to specific chemical phenomena. Some of the earlier chapters include boxes separated from the main text, and which contain further information or fuller details about the subjects covered in the accompanying main text. I suggest that you read the main text of such chapters first, and then venture into the boxes once the basic thrust of the chapter is secure.

I have been helped in the preparation of this book by many people at Basil Blackwell, particularly Romesh Vaitilingam, Mark Allin, Tracy Traynor and Gillian Bromley. Many thanks are due to all of them; and thanks also to my wife, Margaret, for her constant patience, support, and her heroic efforts to divert our infant son from his many attempts to interrupt the work and destroy the word processor and its discs.

1 Prelude

I lie in bed, contemplating chemistry. I am a chemical machine whose parts are made of atoms, molecules and ions. These merge, they mesh, they interact, and in a highly structured sea of seething reactivity they make me. There is no whirring or grinding of the machinery to notify me of the frantic microworld which sustains me. I am aware only of the few calm thoughts within my head, of the gentle ventilation of my lungs, and, if I concentrate very hard, of the rhythmic background pulsing of the blood which nourishes and cleans every corner of my being.

I keep my eyes shut, aware of the chemical complexity waiting to flicker into life if I should open them. Instead, I think downwards. I think of the fabrics of my nightclothes and sheet and mattress cover— long chains of molecules aligned and intertwined, first bearers of the burden of the sleepy 90 kilogram machine that lies above them. Then I think of the coiled and straining metal of the mattress springs, holding me 30 metres from the ground thanks to the chemical tension within them, thanks, ultimately, to the attractions and repulsions between dancing electrical charges. Below the springs, the fabric of the mattress again, then the hard metal that holds it up thanks to the forces that bind together countless metal atoms. These rest on the once living but now long dead chemistry of wooden legs, which rest on the fabric of the carpet, then the chemistry of wood again, and then thick layers of the cold chemistry of stone blocks whose atoms and ions feel my weight all the way down to the ground. Even at the ground itself, my 90 kilograms still have their effect, imposing strains upon the chemicals of the earth and all its sands and stones and liquors down to the hot and sloshing sea of molten iron and its solid core 3, 900 miles and 30 metres, approximately, below me.

I am chemistry, I am sustained by chemistry, my thoughts are chemistry and chemistry supports me on this massive spinning chemical ball throughout its long journey in space and time.

1

At last, I open my eyes and in an instant the magical chemistry of retina and nerve and mind bursts into colourful activity. The light of the sun is penetrating through the filamentous chemicals of the curtains, to fall upon a wall where patterned pigments absorb and reflect according to automatic chemical law to bathe my eye in reds and yellows and greens and blues. Chemistry creates the impression of flowers, the chemistry of my mind performs its astonishing feat of recognition once more; and then some deeper and more subtle mental chemistry reminds me, yet again, that I do not like the pattern.

Suddenly, I perform the seemingly magical feat of movement. The chemical reactions in my mind allow me to decide it is time to turn and observe the clock, and sure enough my head twists across the pillow in obedience. It allows the silvery light from the hands to burst onto my retina and in an instant the chemistry of my brain reacts and recognizes. It is nine thirty, it somehow informs me. I have overslept, again.

From the street below there comes the rattle and roar of the traffic—a constant stream of vehicles fashioned from chemicals won from the earth or created by its most adventurous animal; and powered by the rapid bang, bang, bang, of exploding petrol in smooth metal cylinders. The power of chemical reaction is sustaining my life, cradling me 30 metres from the ground, powering the traffic below me and thundering within the engines of the jet plane which I suddenly hear far above me. Once the plane has passed, I hear another sound—a gust of wind disturbing the transparent chemistry of my window pane. I remember the air and the weather and its cycles of wind and rain—all chemistry again.

There is no getting away from it—chemistry is everywhere within me and all around me. I try to think of things that do not depend on chemistry—anything—but I fail. My mind, my body, my home and all its materials and contraptions and stores of foods and drinks and drugs, the city around me and the planet's land and sea and air . . . It is all chemistry, all part of the massive global, universal, chemical reactor. My mind turns to the people who have made it possible for me to think these things—the chemists who have learned the secrets of, and then harnessed and exploited, the natural power of the chemical world.

There have been chemists around ever since mankind emerged as a species distinct from the other creatures of the earth. Some of the first of these chemists were the people who watched the natural bush and forest fires of the world and learned how to use and purposely generate the consuming chemistry of flame. These first chemists gave humans fires with which to keep themselves warm and simple torches to light the darkness and perhaps ward off fierce animals; but they also offered ways of using fire to torture and execute one's enemies and burn their dwellings to the ground. So the discoveries of the first chemists mirrored the discoveries of all subsequent chemists, in making the natural power of chemistry available as a tool both for good and for harm.

Through the ages chemical knowledge has allowed us to live better, but also to kill better. We could compile a long list of corresponding 'good' and 'evil' purposes to which chemical knowledge has been put: fires for warmth or devastation; explosives for quarrying or killing; new biologically active chemicals for use as medicines or as poisons; fuels for missiles or for cars, and so on . . . Chemical knowledge, like all scientific knowledge, gives us new powers over the forces of the world; but the use to which we put those powers is up to us.

Amongst the good uses to which the first fires could be put were the chemical transformation of meats and vegetables by the heating process we know as cooking, and the initiation of chemical reactions which are able to release metals such as iron and copper from their ores.

The first pure drugs and medicines were won by chemists from within the natural plants and substances around them, while nowadays many are manufactured anew according to prior specification.

Chemists learned how to modify natural products such as cotton to generate new fabrics such as rayon; and then they learned how to make the completely unnatural fibres and plastics so common in the modern world.

Other chemists learned how to convert the nitrogen of the air into the ammonia which can be used as a fertilizer for our crops, while others discovered how the light that reflects from our faces can be captured and focused to form the recognizable chemical imprints we know as photographs.

Chemists have discovered and purified materials which conduct electricity readily, and others which do not conduct it at all, and still others which form the semiconductors which made the transistor and microchip revolutions possible. Taken together, these various chemicals are fashioned into all of the electrical equipment on which the modern world depends.

Chemists have become able to harness the free energy of the sun and convert it into useable heat and electricity, and they continue to strive to perfect and improve these systems in a way which might eventually allow us to trap all of our power directly from the sun.

Evidence of the cleverness of chemists is all around us, and every hour of every day the chemists are busily striving to create new chemicals and get them to do new things which will help us to live better, longer and happier lives; or, of course, to kill and torture one another more efficiently and in ever greater numbers, or perhaps slowly to poison the world with the pollution which the exploitation of our chemical knowledge can create.

Whether you concentrate on its uses for good, or bad, or ugly, chemistry is central to everything around us. Without some understanding of chemistry you can never hope to understand the world.

2   Models in the mind

A lot of people believe that science tells us the way things are, and why they are, but they are mistaken. Science tells us how things *seem* to be. It provides descriptions of the universe which are really only imperfect 'models' which help us to predict and exploit the world around us, but which can never correspond exactly to the way things really are. They are called models because they are incomplete and often simplistic representations of the truth, rather than the truth itself. A scientific theory or model is something which exists only in the human mind, although it corresponds in some acceptably approximate way to the true reality which that mind is trying to comprehend.

Perhaps I am slow on the uptake but throughout my early education, until my first year at university, I wondered why science teachers and lecturers described what they taught as 'theory' rather than 'fact'. On entering my first undergraduate lectures on 'chemical theory', for example, I expected to be told the facts of chemistry, perhaps with a few of the most interesting modern theories thrown in, but surely not theory alone. I soon realized, as the trail of scientific thought became plain to me, that I had been hoping for too much. The ideas which one generation is tempted to describe as facts consistently become discarded theories to the next. I had to accept that the modern ideas being presented to me on the blackboard would, in all likelihood, suffer the same fate. Some of them already have.

This does not mean that what I was taught was nonsense, or that what you will read in this book is nonsense, or that what chemists of a hundred years ago or two hundred years ago thought was nonsense. It simply means that scientific ideas develop from 'rough and ready' approximate descriptions of the universe, to ever more intricate and accurate ones which are nevertheless still likely to be incomplete. Even the rough and ready ideas are useful, allowing the behaviour of the universe to be described, predicted and exploited more effectively than it could be without the help of the theories; and as time passes the utility of the prevailing theories increases as they become more sophisticated but they all prove to be mere theories all the same—incomplete and perhaps incompletable models of reality rather than reality itself.

Thousands of years ago, when people first pondered on the nature of the world, they decided that either material objects must be made up of many individual, distinct and indivisible particles, that matter must be 'discrete' in other words, or else all objects must somehow be woven from some sort of 'continuous' universal fabric. The Greeks, most notably Democritus, decided that matter is discrete and composed of tiny indivisible particles called atoms. This theory allowed John Dalton, early in the nineteenth century, to make sense of all the chemical reactions then known. He described chemistry in terms of the re-arrangement of many types of indivisible atoms, with these atoms combining in fixed proportions to form many other more complex substances known as compounds. And yet the atoms of Dalton's theories are not indivisible. We can now split them up into smaller sub-atomic particles called protons, neutrons and electrons, and the electrons are actually exchanged between atoms in many of the chemical reactions studied by Dalton. So Dalton's insights into the nature of chemistry were based on a false theory. The atoms he studied were not indivisible and were in fact exchanging little bits of themselves all the time right under his nose. Yet his insights were still very valuable, and remain valuable and valid today. Much of chemistry really does consist of different atoms combining in fixed proportions to form more complex compounds, but nowadays we do not regard the atoms as indivisible.

So when you read about the modern ideas of chemistry throughout this book, you should not be seduced into faithfully believing that 'this is the way chemistry *is*'. Instead, I will be telling you of the way chemistry seems to be and the way chemicals seem to behave, using the theories of today which, despite their great power, remain mere theories and models all the same. Just as Dalton was able to gain insight into the world by realizing that in many ways it behaves as if it consisted of indivisible atoms, although it does not, so modern chemists gain insight into the world using theories and models of reality that may well be incomplete or untrue. In fact, as you will see, modern chemists sometimes use two or more different and seemingly conflicting theories to help them understand the one phenomenon, recognizing that neither theory is perfect, but that each has some use as an approximation to reality in some circumstances. A scientific theory is a way of regarding the world, and there are several different ways of regarding the world of chemistry, all of which are useful approximations to the deeper and largely hidden ultimate truth (see figure 2.1).

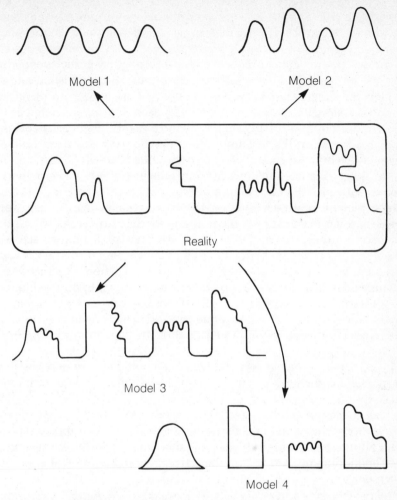

Figure 2.1   Scientific theories construct 'models' of reality to varying degrees of accuracy and utility. Different and seemingly conflicting models can all contain useful elements of truth.

So if you think that learning about chemistry, or any other branch of science, involves simply learning a series of known facts, and a series of mysteries yet to be solved, think again! Science is a far less certain and more subtle business than that, with even the seemingly most secure and fundamental aspects of scientific theory always remaining uncertain and open to challenge. For example, most people think that Democritus pinned down the general nature of matter once and for all over 2,000 years ago, when he decided matter was composed of discrete fundamental particles rather than some sort of continuous fabric; and yet some aspects of modern physics suggest a very different picture. They describe the entire universe and all its individual particles as mere 'knots' or 'entanglements' of the continuous and indivisible fabric which physicists refer to as 'space-time'. So although the physical world certainly behaves *as if* it were composed of many discrete particles, and although all of the persuasive and powerful chemical theory of today is based on that assumption, at the deepest and most fundamental level matter may really be continuous after all.

Working chemists and physicists and biologists rarely concern themselves with such dilemmas. They are content to study and understand and exploit the natural world from the viewpoint of whatever theory or model proves most useful at the time and for the purposes they desire. If insurmountable obstacles or inconsistencies present themselves, then the theory must change (as it did when atoms turned out to be divisible into smaller sub-atomic particles); but when a theory does change all of the old facts are readily incorporated into it in a slightly modified form. Changes to scientific theory, even apparent revolutions in our understanding, tend to refine and add to our previous picture of the world, rather than wiping it out altogether.

My cautionary preamble is complete. I can now begin to describe the chemical microworld without constantly reminding you that the description relies on theories and models of reality that are open to future challenge and change. These theories and models allow us to construct a fascinating tale about the unseen world within ourselves and all the objects around us, a tale which gives us the insight and understanding needed to turn the activities of that inner world to our own ends.

3   Into the atom

Everything we see happening around us, and in the universe at large, involves chemicals; so to investigate the nature of chemistry we merely have to take a look at the universe, particularly the bit of it we inhabit, which we call the world. What do we see if we look at the world in search of some glimpse of its real nature?

Our first observation is that there are lots of 'things' in the world, things that sometimes change. In fact, if we look very closely, we see that all things are always changing. 'There are things and they change' is not a very complete or useful description of the universe, but it is a first rough and ready one which does have some utility despite its brevity. As we look closer, and think deeper, we can begin to classify the things and the changes.

One of the next observations most people would make about the world, if they came into it with the intelligence of an adult and the ignorance of an infant, is that there are hard 'solid' things, and sloshy 'liquid' things, and ghostly 'gassy' things which we can often feel as they waft and blow around us but cannot always see. The labels we give to these things do not matter, but there are clearly three main types of things—solids, liquids and gases.

It is convenient to have a name to describe the stuff that things are made out of, to help distinguish the things from other more subtle phenomena such as changes. We say that things are made out of 'matter', and that helps to organize our thoughts even if we have no idea of what matter is. A common dictionary definition of matter is 'anything that has mass and occupies some space', with the 'mass' of an object being a measure of how large a force we must apply in order to accelerate the object to a given extent. Of course this definition tells us nothing of what matter is really made of, or why it has mass.

Many humans have spent a lot of time looking into matter to try to find out what it really is, and they have made some startling discoveries that suggest a welcome simplicity at the heart of all the things of the world.

One of the easiest discoveries to comprehend and to visualize is that matter behaves as if it were made up of many 'little bits of matter' which we call 'particles'. Matter behaves, in other words, as though it were particulate rather than continuous. We can cut large pieces of matter into smaller bits, and grind it down into fine powders, but eventually we would arrive at tiny indivisible particles of matter which we could not cut up any further. Another simple observation of fundamental importance is that the particles of matter are moving—jostling about chaotically as they bump into and bounce off one another. We shall see that this movement is what makes changes happen. It is what creates the dynamic world around us and allows us to live in it.

It is very easy to see the effects of matter's moving particles by looking at some smoke through a simple microscope. The smoke needs to be contained within a transparent vessel, and have a strong light shining through it. A glance into the microscope reveals many dancing specks of light. The specks correspond to the tiny lumps of ash that make up the smoke, but although they are tiny to us these lumps are enormous compared to the smallest particles of matter. So the dancing specks are not themselves the moving particles of matter we seek evidence of, but they are dancing because of the random bombardment they receive from the much smaller particles of the air. The particles of the air are constantly on the move, rushing around at hundreds of miles per hour to collide with one another and with the great lumps of ash which reflect our light. As we look down the microscope we see the lumps of ash being tossed around within a sea of seething particles. We cannot see the particles, but we can see the violent effects of their existence and their motion.

Our description of the universe has become more sophisticated. From 'there are things and they change', we have moved on to 'there are things which are all composed of tiny moving particles, and the movement of the particles brings about change.'

It is time to become more specific, drawing on the discoveries of many different people to discuss the particles of matter in more detail.

As we look ever deeper into the structure of matter, we find increasing simplicity rather than deepening complexity. The material objects of the earth are infinitely varied and virtually infinite in number, yet, apart from a few exotic exceptions, they are all composed of just three types of particle, known as atoms, ions and molecules. To make things even simpler, the molecules are all composed of atoms joined together, and the ions are either molecules or atoms modified in a very simple way, as we shall see. So atoms are the fundamental particles of chemistry, but they are not 'fundamental particles' in the true sense, since atoms are themselves composed of varied numbers of three 'sub-atomic particles', the 'protons', 'neutrons' and 'electrons'.

To understand all chemistry we need delve no deeper into matter than the level of protons, neutrons and electrons; but it is possible to go deeper and find further simplicity. Protons and neutrons are both composed of fundamental particles known as 'quarks' (three quarks per proton or neutron). There seem to be no deeper structures within electrons, though, so all the material variety of the world is based on different arrangements of just two types of fundamental particles: electrons and quarks. There are actually several different types of quarks, though only one type of electron, but quarks need bother us no further.

Protons and neutrons and electrons are the stuff of chemistry, the matter that matters in chemical reactions. Chemists need to understand atoms, and that is really all they need to understand. The other particles which participate in chemistry are derived from atoms when they interact, or 'react' to one another's presence. The atom is the basic architectural unit, the simplest 'building block', of all the great cathedrals of matter such as you and I and elephants and trees and aeroplanes and the entire earth itself.

The view of atoms which chemists must survey looks both inwards and outwards. It looks inwards to discover the inner nature of atoms, which holds them together and makes them work; and it looks outwards to examine what happens when atoms bump into one another. Everything that you would call chemistry is encompassed by this view. We shall look inwards first.

We must examine three things—protons, neutrons and electrons—and try to understand what they are and how they behave. Thankfully, there are only two essential properties of these things for us to consider: their 'mass' and their 'electrical charge'.

The mass of something is just a measure of the amount of matter it is made up of. In loose terms it is a measure of how heavy something is. All masses are relative—we can only quote the mass of something relative to something else, just as we can only quote the weight of something relative to some standard weight, such as a kilogram or a pound. So we can say something has twice the mass of something else, meaning it contains twice as much matter, or has a thousand times the mass, or one ten thousandth of the mass of something else; but there are no absolute units of mass, it is a relative term. Conveniently, scientists have chosen units of mass which make the mass of a proton very nearly one 'atomic mass unit' (1 amu). It is not exactly 1 amu, because of the precise way in which the amu is defined, but it is very close to it. You can get some impression of how tiny this mass is if I tell you that, converted into the more familiar units of kilograms, it becomes 1.67×10^{-27} kg (which is a mathematician's shorthand way of representing 0.00000000000000000000000000167 kg). So compared to the sort of objects we are familiar with a proton has an incredibly small mass. It is also incredibly small. The diameter of a proton is 1×10^{-15} metres (0.000000000000001 metres), so it is very small and very light.

The mass of a neutron is also very close to 1 amu, and it is roughly the same size as a proton as well. There are small differences between the masses and sizes of protons and neutrons, but too small for us to have to bother about. We can usually consider protons and neutrons as tiny particles of matter with much the same mass (1 amu) and very similar sizes.

Although protons and neutrons are tiny compared to us, they are very large compared to electrons. The mass of an electron is 9.1083×10^{-31} kg (in other words 0.00000000000000000000000000000091083 kg), which in more meaningful terms is 0.000545 times the mass of a proton. This means that the combined mass of 1,833 electrons is equal to the mass of just one proton. Electrons contain so little matter they are hardly there at all, yet we shall see that they are responsible for all the richness and diversity of the chemical world.

The really crucial difference between protons, neutrons and electrons, the difference which lies at the heart of all chemical change, concerns the type of electric charge they carry. To investigate this, we must first consider what electric charge is. I could say 'nobody knows for sure', and be very close to the truth, although that would not help us. Electric charge is the name scientists have given to a phenomenon which they can describe in great detail but not truly explain.

The process of making sense of the universe involves observing what happens in it and trying to find links between various phenomena which allow us to reduce complex events and phenomena into a pattern of simpler events and phenomena. Lots of objects in the universe appear to be 'pulled' or 'attracted' towards some other objects, and repelled from others, in a manner that suggests the same force or agency is involved. The force has been called the electric force, and objects which feel the effects of this force are said to carry electric charge. These are man made labels with no real meaning in themselves. Completely different labels could be chosen, but the phenomena they describe would remain the same.

It is easy to see some of the effects of these phenomena. When you quickly pull a nylon shirt over your head, for example, you sometimes hear a crackling noise and discover that a few of your hairs are sticking out, or even standing on end. A scientist would say your hairs are behaving like this because they have acquired an electric charge 'of the same sign'. Scientists have found it makes sense to assume there are two signs of electric charge: 'positive' (+) and 'negative' (−). Objects

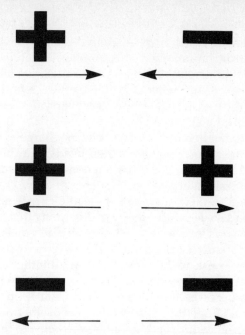

Figure 3.1 The electric force causes charges of opposite sign to be attracted towards one another, and charges of like sign to be repelled from one another.

carrying charges of the same sign (either positive or negative) are repelled from one another, hence the bizarre behaviour of your hair. Objects carrying charges of opposite sign (one carrying a positive charge and the other a negative charge) are attracted towards one another (see figure 3.1).

We could have invented all this nomenclature simply in order to try to describe the behaviour of hair and nylon shirts. We would notice that the hair is attracted towards the shirt whose pulling caused it to acquire its charge. So the hair and the shirt, in other words, seem to carry opposite charges. This might lead us to suppose that something carrying electric charge of one sign has actually been transferred between the hair and the shirt, leaving one of these objects deficient in one type of charge, while the other gains a surplus. We might easily reason that both objects began with an equal mixture of positive and negative charge, but as a result of the pulling a quantity of one type of charge has been 'rubbed' onto the other object, leaving one positively charged overall, the other negatively charged overall, and causing them to be attracted towards one another. This reasoning would also explain why the hairs are repelled from one another, because they have acquired either an excess or a deficiency of one type of charge, causing them all to be left carrying an overall charge of the same sign, and hence to be repelled from one another. This simple reasoning process reveals all the essential principles about why your hairs can stick out and be attracted to a nylon shirt when it is pulled off over your head. It is, in fact, all due to the transfer of electrons, each carrying a negative charge, from the hair to the shirt.

Scientists did not invent the idea of electric charge by looking at the effects of pulling nylon shirts over their heads, but they could have. They invented it by looking at other phenomena, including the effects of rubbing natural materials such as amber and glass, and they soon became so used to their neat idea of electric charge and the electric force, and found it such a useful way of describing things, that most of them stopped worrying about what the terms really meant. They found that their ideas about electric charge and the electric force allowed many different phenomena to be related to one another in a way that suggested that what they had described as electric charge and the electric force must lie behind them all. When something which begins as a description of one thing turns out to be useful in describing and predicting the behaviour of other things, you begin to realize that your invented description must be a pretty close match to the hidden reality of the universe, whatever that is. That is what happened to the invented ideas of electric charge and the electric force.

When scientists began to build apparatus that could probe into the structure of atoms, they found electric charge at work in there as well. In fact they identified what seemed to be the fundamental source of all the other phenomena which we would describe as being 'electric' in origin, for they found evidence of two types of particle which carried electric charge *of equal amounts but opposite signs*. These particles are the proton, which carries a positive electric charge of $+1$, and the electron, with a negative charge of -1. (We could call the charge on the proton negative, and the charge on the electron positive, if we wished, and nothing would change. The terms themselves, like all terms, have no inherent meaning, they are merely words we choose to describe the way things are, or seem to be.) As I have already said, the electrons are the charge carrying particles which are transferred from hair to shirt when you pull a nylon shirt off over your head, creating the charge imbalances that make your hairs repel one another and become attracted towards the shirt.

So, in our system of scientific language, in which experience has shown it makes sense to describe some phenomena in terms of electric charges and the electric force between them, protons appear to be basic carriers of positive charge, while electrons are basic carriers of negative charge. This means protons will be repelled from one another (unless some other effect holds them together) and attracted towards electrons (unless some other effect resists this attraction). Equally, electrons will be repelled from one another (unless some other effect holds them together) and attracted towards protons (unless some other effect resists this attraction).

These forces of attraction and repulsion are the forces that make chemical reactions happen. All of chemistry can be reduced to a frantic electric dance of whirling electrons and protons; and the electrons are the dancers which move the most as they are pushed and pulled from place to place like the most sought after partners in a dance hall. The story of chemistry is essentially a story of electron rearrangement, as we shall see.

I did not mention the neutron in the discussion of electric charge, because it carries no overall charge. It is, in other words, electrically 'neutral'; hence the name. That fact leaves our description of the three sub-atomic particles complete.

Earth, water, air and fire were once considered the four most basic 'elements' of the world, but this early attempt to identify the simplicity at the heart of nature's complexity was misguided. Earth, water, air and fire are chemically complex things composed of different combinations of the true elements. Modern chemists recognize 92 basic elements of the earth, with an element being defined as a substance composed of only one kind of atom. So if there are 92 different chemical elements on and in the earth, there must be 92 different kinds of atom. All atoms, however, are made up of different arrangements of protons, neutrons and electrons, so the hierarchy of chemical simplicity begins with a trio of sub-atomic particles, which combine to form the 92 types of atom which occur naturally on earth, which themselves combine to form the almost infinite variety of other chemicals all around us. These other chemicals are known as 'compounds' since they contain atoms of several different elements joined together in ways we shall consider in chapter 6.

The 'periodic table of the elements' (see figure 3.3) lists the atoms of the world beginning with the simplest, hydrogen (H), and moving through helium (He), lithium (Li), beryllium (Be), boron (B), carbon (C) and so on, until reaching the largest and most complex atoms near the bottom. (To move steadily from the smallest through to the largest atoms you read along each row of the periodic table from left to right, and progress through the rows from top to bottom, apart from where instructed by an arrow to jump to a new location). There are 108 elements in the periodic table as shown in figure 3.3, rather than only

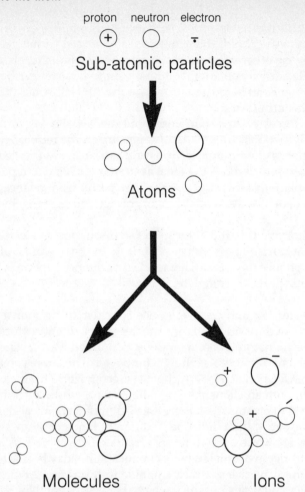

Figure 3.2 The sub-atomic particles combine to form atoms, which can combine to form molecules. Ions are formed when atoms or molecules lose or gain electrons to acquire a net electric charge. Protons, neutrons, electrons, atoms, molecules and ions are the basic particles of matter of the chemical world.

92, because scientists have succeeded in artificially manufacturing 16 elements which are larger and more complex than any found naturally on earth. Before exploring such achievements (which is done in box 3.1) we must investigate the structure of atoms, and the differences between different types of atom. The best place to begin is with the simplest type of atom, the hydrogen atom.

A hydrogen atom is a very simple thing indeed. It is a proton surrounded by an electron (see the top atom of figure 3.4). The proton, of course, carries a charge of $+1$, while the electron carries a charge of -1, so the hydrogen atom, like all atoms, is electrically neutral overall. Looking at the picture of a hydrogen atom in figure 3.4 presents us with a puzzle. Why is the atom so big? We can see the proton, at the centre of the atom, but the electron, despite being tiny compared to the proton, appears to occupy a much larger volume of space. Why is the electron not simply nestling up against the proton? After all, it is attracted towards the proton by the electric force.

The traditional resolution of the dilemma was to say that the electron must be moving. It must be whirling around the proton a bit like a satellite in orbit around the earth. Anything that is moving is said to possess some 'energy', an idea whose meaning will be fully explored in chapter 4. So the electron of a hydrogen atom does not nestle into the proton, but instead it occupies a large region of space around the proton, because it has the energy needed to defy the electric force which would otherwise pull the electron into the proton. It used to be thought that the electron really did 'orbit' around the proton like a satellite orbiting the earth. It soon became clear, however, that this was a misleading picture. All that a modern chemist will say about the electron is that, at any instant, it is 'probably' somewhere within the spherical 'orbital' shown in figure 3.4. The chemist could calculate for you the precise probability of the electron being at any particular place within its orbital, and would point out that there is even a possibility that the electron might lie outside of the orbital boundary shown in the figure. This boundary is drawn to give roughly a 90 per cent probability of finding the electron somewhere within it. So the probability of finding the electron outside the boundary is 10 per cent overall, and steadily diminishes the farther out you go. If you demanded the right to visualize the movement of the electron, the chemist would tell you to imagine it darting about erratically within the spherical orbital shown in the figure, and making occasional forays outside it, but he would point out that modern theories lead us to question whether we should regard the electron as a tiny discrete particle at all. They suggest instead that it should be considered as being somehow 'smeared' throughout the entire volume of its orbital (for reasons discussed in chapter 4).

Remember that chemical theories represent models of chemical reality which presumably draw closer to the truth as they become more precise and sophisticated. Even the most crude and simplistic models are useful; indeed, it is a mistake to insist on the most rigorous model possible if a more approximate one will meet your needs. You should work with the model that best meets your needs and most facilitates understanding. In this book we are simply trying to appreciate what

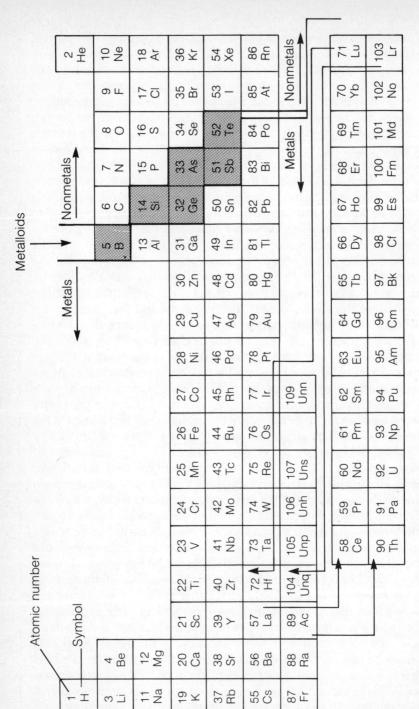

Figure 3. 3(a)   The periodic table of the elements.

Name	Symbol	Atomic number	Name	Symbol	Atomic number
Actinium	Ac	89	Neodymium	Nd	60
Aluminum	Al	13	Neon	Ne	10
Americium	Am	95	Neptunium	Np	93
Antimony	Sb	51	Nickel	Ni	28
Argon	Ar	18	Niobium	Nb	41
Arsenic	As	33	Nitrogen	N	7
Astatine	At	85	Nobelium	No	102
Barium	Ba	56	Osmium	Os	76
Berkelium	Bk	97	Oxygen	O	8
Beryllium	Be	4	Palladium	Pd	46
Bismuth	Bi	83	Phosphorus	P	15
Boron	B	5	Platinum	Pt	78
Bromine	Br	35	Plutonium	Pu	94
Cadmium	Cd	48	Polonium	Po	84
Calcium	Ca	20	Potassium	K	19
Californium	Cf	98	Praseodymium	Pr	59
Carbon	C	6	Promethium	Pm	61
Cerium	Ce	58	Protactinium	Pa	91
Cesium	Cs	55	Radium	Ra	88
Chlorine	Cl	17	Radon	Rn	86
Chromium	Cr	24	Rhenium	Re	75
Cobalt	Co	27	Rhodium	Rh	45
Copper	Cu	29	Rubidium	Rb	37
Curium	Cm	96	Ruthenium	Ru	44
Dysprosium	Dy	66	Samarium	Sm	62
Einsteinium	Es	99	Scandium	Sc	21
Erbium	Er	68	Selenium	Se	34
Europium	Eu	63	Silicon	Si	14
Fermium	Fm	100	Silver	Ag	47
Fluorine	F	9	Sodium	Na	11
Francium	Fr	87	Strontium	Sr	38
Gadolinium	Gd	64	Sulphur	S	16
Gallium	Ga	31	Tantalum	Ta	73
Germanium	Ge	32	Technetium	Tc	43
Gold	Au	79	Tellurium	Te	52
Hafnium	Hf	72	Terbium	Tb	65
Helium	He	2	Thallium	Tl	81
Holmium	Ho	67	Thorium	Th	90
Hydrogen	H	1	Thulium	Tm	69
Indium	In	49	Tin	Sn	50
Iodine	I	53	Titanium	Ti	22
Iridium	Ir	77	Tungsten	W	74
Iron	Fe	26	Unnilnonium	Unn	109
Krypton	Kr	36	Unnilpentium	Unp	105
Lanthanum	La	57	Unnilhexium	Unh	106
Lawrencium	Lw	103	Unnilquadium	Unq	104
Lead	Pb	82	Unnilseptium	Uns	107
Lithium	Li	3	Uranium	U	92
Lutetium	Lu	71	Vanadium	V	23
Magnesium	Mg	12	Xenon	Xe	54
Manganese	Mn	25	Ytterbium	Yb	70
Mendelevium	Md	101	Yttrium	Y	39
Mercury	Hg	80	Zinc	Zn	30
Molybdenum	Mo	42	Zirconium	Zr	40

Figure 3.3(b)   The elements' names, symbols and atomic numbers.

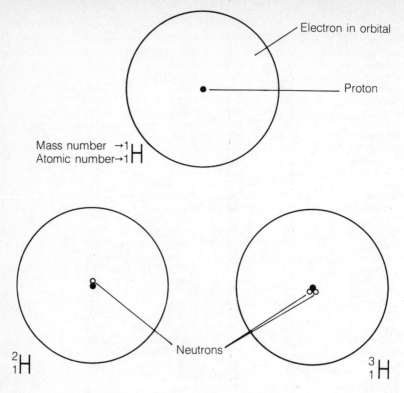

Figure 3.4 The isotopes of hydrogen.

chemicals are and why chemical reactions happen, and for these purposes the image of an electron as a tiny particle of matter, carrying a negative charge, and rushing about erratically within its orbital is perfectly acceptable. It is a close and very useful approximation to the truth.

We are going to have to return to the question of the energy that electrons must possess in order to occupy their orbitals within atoms, but before we do we should consider the architecture of atoms some more.

If you were able to shrink down to the dimensions of the chemical microworld and examine a series of hydrogen atoms, you would discover something a bit different about a few of them. In some, the proton at the heart of the atom would be bound to a neutron. In others, the proton would be bound to two neutrons. These two variant types of hydrogen atom are very rare; you might find one or two of them for every 10,000 of the normal type, but they do exist and they are important. They illustrate the general point that not all the atoms of any particular element, such as hydrogen, are identical, because atoms of the same element can have different numbers of neutrons present within them. The presence or absence of neutrons does not alter the basic chemical character of atoms. That basic character, governing what reactions the atom can participate in, is determined by the number of protons and electrons it contains. Chemistry, as we shall see, is all about the interaction between positive and negative electric charge, and since neutrons are electrically neutral, they have no real influence on the chemical nature of the atoms that contain them (apart from some slight and subtle effects).

All hydrogen atoms contain only one proton, and any atom containing only one proton must be an atom of hydrogen. The number of protons in an atom is what determines what type of atom it is. In recognition of this importance the number of protons in an atom is known as its 'atomic number'. So the atomic number of an atom defines what type of atom it is. Since all atoms are electrically neutral overall they must all contain the same number of electrons as of protons, but the proton number is the more fundamental characteristic. When atoms react they can sometimes lose or gain electrons, as we shall see, but the number of protons they contain never changes during the course of a chemical reaction. There are situations in which the number of protons in an atom will change, but when this happens the atom is transformed into an atom of a different element (as discussed in box 3.1).

So the basic principle of atomic architecture is as follows: all atoms of any one type (of any element, in other words), contain the same number of protons and electrons, and a variable number of neutrons.

The variability in the number of neutrons means that atoms of any element come in several varieties which differ simply in the number of neutrons they contain. These varieties are known as 'isotopes' of the element. So three isotopes of hydrogen exist in nature—the one with no neutrons (and so with a total mass of 1 amu), the one with one neutron (total mass = 2 amu) and the one with two neutrons (total mass = 3 amu). These three forms, together with a shorthand way of representing them, are shown in figure 3.4. Remember that the mass of an atom's electrons is always negligible, compared to the mass of its protons and neutrons, so in calculating the total mass of atoms we can usually ignore the electrons.

There is one more term to be introduced before we look at some more complex atoms. All of the protons and neutrons of an atom are bundled together into a tiny central core known as the atomic 'nucleus'. This nucleus is always tiny compared to the much larger volume in which the electrons are found, although it carries virtually all of an atom's mass. The nucleus is like the massive but compact earth which is orbited by many tiny electron 'satellites' which are found throughout a much larger volume of space.

We have looked at the way in which a proton, an electron and between zero and two neutrons combine to create atoms of hydrogen, the simplest element. Now we must examine a few larger and more complex types of atom.

Having been told that hydrogen is the simplest element, each hydrogen atom having just one proton in its nucleus and therefore an atomic number of 1, you should not be surprised to learn that atoms of the next most complex element have two protons in their nucleus and an atomic number of 2. These are called 'helium' atoms. All helium atoms contain two protons, and therefore two electrons, and between one and four neutrons (usually two).

Looking at a helium atom (figure 3.5) presents us with a new puzzle. It shows that each atom contains two protons bound together in the nucleus, yet we know that these protons should be repelled from one another by the electric force which pushes apart objects carrying charges of the same sign. Since both protons carry a charge of +1 why do they not fly apart? What keeps them bound together within the nucleus? The answer is another force, different from the electric force and stronger than it over short distances.

The electric force is not the only force at work in the universe. Scientists actually recognize four 'fundamental forces' which they believe are responsible for all of the pushing and pulling and changing which make things happen. One of these fundamental forces is the electric force,* felt by objects carrying electric charge; another is the familiar force of gravity, which attracts all objects with mass towards all other objects with mass; and a third is the 'strong nuclear force' which binds protons and neutrons into the atomic nucleus, despite the counter-efforts of the electric force to break the nucleus apart. So the fundamental forces of nature need not act in harmony, but can act in opposition to one another, with one force proving dominant in one situation, while another may prove dominant in other situations. The

---

*Strictly speaking we should say the 'electromagnetic force' in place of the 'electric force', since the electric force is only one aspect of the electromagnetic force which, as its name implies, is also responsible for the phenomenon of magnetism and the forces acting between magnetic poles. The term electric force is commonly used, however, when referring to only the electric effects of the electromagnetic force—its simple pushing and pulling effects on positive and negative charges, in other words.

strong nuclear force, for example, is only strong over very short distances. So when protons are very close, as in the atomic nucleus, it can overcome the repulsive effect of the electromagnetic force; but if we could somehow prize the protons of the helium nucleus, for example, a little way apart, then the strength of the strong nuclear force would fall dramatically, the electric force would become dominant, and the protons would fly apart.

The fourth fundamental force is called the 'weak nuclear force'. As the name suggests it is rather weak and acts within atomic nuclei, where it is responsible for some subtle effects which we need not consider here.

The *reason* for the existence of the four fundamental forces is a problem for physics, not chemistry. Physicists have some fascinating ideas about the origins of and relationships between these forces, which you can find out about by reading the physics books listed in the Further reading section. Chemists merely need to accept that these forces exist, and get on with the interesting and very useful task of discovering how and why they make atoms do the things they do.

As we climb the ladder of atomic complexity, from hydrogen (atomic number = 1), to helium (atomic number = 2) , and then onwards, the next most complex atom is 'lithium', with an atomic number of 3. This means that lithium atoms contain three protons and three electrons, and most also contain four neutrons (although, as always, the neutron number can vary between different isotopes of the element).

The diagram of a lithium atom (see again figure 3.5) introduces a new and vital principle of atomic architecture. It shows that two of the atom's electrons are in an inner orbital, much the same size as the orbital which holds the two electrons of a helium atom, or the one electron of a hydrogen atom (a shaded orbital in the figure represents an orbital containing two electrons, while an unshaded orbital contains only one electron). The third electron of the lithium atom, however, is in a different, larger orbital. Why is that?

Remember that an electron orbital is simply a volume of space in which an electron is likely to be found. It is only possible, however, to 'squeeze' two electrons into any one orbital. This situation is analagous to the problem facing satellite users as they place ever more satellites into the increasingly crowded space around the earth. If they put too many satellites into the same orbit, they may begin to collide with one another, or at least the risk of collision will become unacceptably high. This is only an analogy for the situation of electrons, but it helps us to appreciate the reason why there is a limit to the number of electrons that can occupy each orbital. The analogy breaks down rather when we realize that the two orbitals shown in the lithium atom *overlap* with one

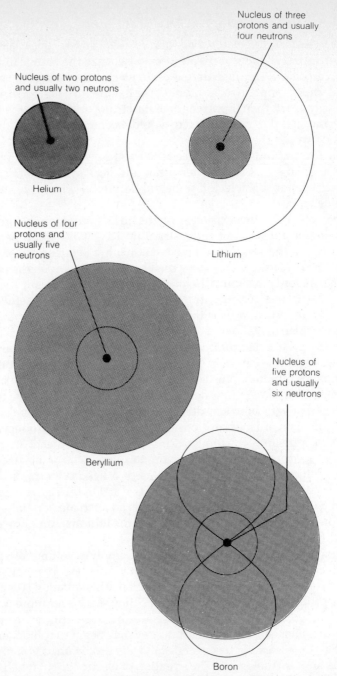

Figure 3. 5 *Some atoms and their occupied electron orbitals. Shaded orbitals are 'full', i.e. they contain two electrons; unshaded orbitals contain only one electron. The dumb-bell-shaped orbital of boron contains only one electron, although this is largely obscured by orbital overlap.*

another. They are simply two spherical volumes of space around the atom's nucleus, one bigger than and incorporating the other. The electron in the outer orbital can be found anywhere within that orbital, including the central region which is also occupied by the inner orbital. The presence of the outer orbital clearly increases the total space available to and occupied by the three electrons, so the idea of larger orbitals providing the extra room needed for atoms to accommodate larger numbers of electrons, although an imperfect analogy, is still a helpful one as we try to visualize the principles involved.

After lithium, we move on to 'beryllium', the atom with an atomic number of 4, and therefore comprising a nucleus of four protons (and usually five neutrons) surrounded by four electrons. Remember that there is room in each electron orbital for two electrons, so the fourth electron fits into an outer orbital similar to that used by the outer electron of a lithium atom. The heavy shading of the outer orbital indicates that this is a 'filled' orbital, containing two electrons, rather than a 'half filled' orbital containing only one electron.

By now all the main the principles of atomic architecture should be becoming clear to you:

Atoms are composed of protons, neutrons and electrons.

The number of protons in an atom always equals the number of electrons, making every atom electrically neutral overall.

Atoms of any particular type, of any element in other words, all contain the same number of protons, and therefore also the same number of electrons, although they may contain different numbers of neutrons.

The electrons of atoms occupy volumes of space known as 'orbitals', each orbital containing a maximum of two electrons.

The electrons tend to occupy orbitals which are as near as possible to the nucleus, i.e. which have the minimum volume.

That last principle has not been explicitly stated before, but should be obvious by looking at the orbitals occupied by the electrons of the four types of atom we have considered so far. You will discover in chapter 4 that the correct way of stating this principle is to say that electrons occupy orbitals of the lowest possible *energy*, but I am leaving the full discussion of energy until that chapter.

If you examine the periodic table (figure 3.3) you will find elements corresponding to all atomic numbers from 1 to 92, the elements that occur naturally on earth, and then continuing with elements manufactured by mankind up to an atomic number of 109. There are no gaps in the ladder of natural atomic complexity—there are atoms with every possible atomic number from 1 up to 92, corresponding to all the atoms found naturally on earth. The unbroken series continues through to 107 due to the ingenuity of mankind, although element number 108 has not yet been manufactured while number 109 has.

There is no point in drawing out the structure of all of these atoms, since the principles involved should be clear to you by now. If I asked you to draw a picture of an iron atom, and in particular the most common isotope which contains 30 neutrons, you would simply need to draw 26 protons and 30 neutrons all bound together to form the atom's nucleus; but what about the electrons? Based on what I have told you so far, you would probably fit the 26 electrons into 13 separate orbitals of increasing volume and all with the same neat spherical shape as the orbitals you have seen so far. Things are not that simple, however. As we investigate atoms beyond beryllium in the periodic table, we begin to find some electrons in differently shaped orbitals. Boron, for example, is the element with an atomic number of 5, and its outer electron goes into a twin lobed or 'dumb-bell shaped' orbital, as shown in figure 3.5.

Other larger atoms have some of their electrons in more complicated orbitals with four lobes or even eight lobes, rather than the two lobes of the outer orbital of boron, with all the lobes projected outwards from the central nucleus. Regardless of their sometimes bizarre shapes, however, the principle governing electron orbitals is very simple: each orbital is simply a region of space which can be occupied by up to two electrons. By looking at the real shape and nature of some electron orbitals, we can see how approximate and deficient is the analogy between electron orbitals and the orbits followed by satellites as they circle the earth. Electrons have much greater freedom than satellites, since they appear to be able to dart about erratically within volumes of space which sometimes have rather complex shapes. It is true, however, that just as it takes more energy (i.e. more rocket fuel) to put a satellite into a high orbit, so an electron in a 'high' orbital which projects far out into the space around an atom's nucleus needs more energy than an electron in a lower orbital whose boundary lies closer to the nucleus. Energy is such a fundamental idea to chemistry, we can put off its full examination no longer.

Box 3.1 The elements

The virtually infinite number of different chemical substances in the world are all composed of different aggregates and arrangements of the elements listed in the periodic table (figure 3.3). Only 92 of these elements occur naturally, and a further 16 have been made by man, but despite this very restricted range, compared to the many millions of more complex chemicals formed when elements combine, many of the pure elements themselves are some of the most abundant and important substances on the earth.

Glance through figure 3.3(b), listing the full names of the elements of the periodic table, and you will find the names of some very important substances indeed. Iron, aluminium, copper, zinc, titanium, platinum, tin, silver and gold are just a few of the elements known as 'metals', which form much of the physical structure of the materials and machines upon which modern society depends. The 'non metals' on the right of the periodic table include oxygen, the gas we must breathe in order to survive; carbon, which forms the diamonds we covet as jewels and whose hardness is also exploited in industry; chlorine, which is added to swimming pools as a disinfectant; and the gas neon which glows within many types of advertising signs and street lights. In between the metals and non metals there are a few 'metalloids'. These are elements which do not really fit into either the metal or non metal categories, but which share some properties with both. The metalloids include the silicon and germanium which are vital materials in the electronics industry, used to make the 'silicon chips' which lie at the heart of all modern microelectronic circuits. Another important and well known element is the uranium used to provide us with nuclear power (as well as nuclear bombs). So although the pure elements, in other words substances that consist of just one type of atom, are relatively small in number, many are important and useful chemicals in their own right.

Each element is slightly different from all the others in the way it looks and feels and behaves. All such differences can be related back to the different structures of the atoms of the elements, in other words to the different assemblies of protons and neutrons and electrons which the atoms consist of.

Although each element is unique, the periodic table can be divided up in various ways into elements which share certain properties. Most are solids at everyday 'room' temperatures, for example, but several exist as gases (including hydrogen, oxygen, nitrogen, helium and neon), while two are liquids (mercury and bromine). Elements in the same group of the periodic table (the same vertical column, in other words) tend to share broadly similar chemical characteristics, so the division of elements into groups represents another useful way of dividing them up. A third way, of course, is the division into metals, metalloids and non metals, already considered. This division is made on the basis of several characteristics, and is not clear cut, but one of the most important distinctions between metals and non-metals is that while metals are good conductors of electricity, most non-metals are very poor conductors of electricity. The reason behind this is quite interesting.

Atoms of the elements which we call metals all tend to have one or more outer electrons (see page 45 for a fuller discussion of what this means) which can rather easily escape from one metal atom and drift about for a while before returning or moving onto another atom. This makes it quite sensible to envisage these metals as consisting of atoms which have become positively charged by losing one or more outer electrons, to form what are known as positive 'ions', surrounded by a 'sea' of the lost outer electrons which can move around between the ions. This electron sea within the structure of metals explains why they conduct electricity so well. An electrical current is essentially a flow of electrons from one place to another, and the electron sea within metals provides plenty of relatively free electrons which can form such a flow. The electron sea of a metal wire, for example, will flow towards any region of positive charge and away from any region of negative charge, due to the electric force of attraction between opposite charges and repulsion between like charges. Such flows of electrons though the structure of metals allow the metals to act as electrical conductors in all types of electrical circuits.

The outer electrons of non-metal atoms are generally held more tightly to their atomic nuclei, so they are not free to escape and become part of a current of moving electrons, and so most non-metals are non-conductors of electricity.

As you might expect, the characteristics of the metalloids lie somewhere in between, which is why they include some important 'semiconductors', such as silicon, so vital to the microelectronics industry.

The electron sea within the structure of metals also acts as a sort of 'electric glue' holding any piece of metal together. The escaped outer electrons are electrically attracted to the metal ions left behind, so they tend to hold these ions together, and hence the whole structure of any piece of metal together. This phenomenon is known as 'metallic bonding' since it bonds the individual particles of a metal into one cohesive structure. The atoms of most non metal elements also tend to be held together into clusters of from a few to a great many atoms. The bonds between such non metal atoms are different, however; they are normally the so-called 'covalent' bonds which you will learn about in chapter 6.

I hope I have convinced you that the 92 naturally occurring elements of the earth include a wide variety of very important and interesting substances, but what about the 'man-made' elements? How can we seemingly play at being God and create new elements?

Recall that what makes each element unique is the number of protons in its nucleus. Uranium atoms, the most complex naturally occurring atoms, have 92 protons in their nuclei; if we could somehow add an extra proton to a uranium nucleus we would create the nucleus of a new element. This can be achieved by firing a beam of fast moving neutrons at a sample of uranium. Neutrons can behave as if they were composed of a proton bound to an electron. When a neutron collides with the nucleus of a uranium atom it can yield up a new proton which can become part of the nucleus, and an electron which is ejected away from the site of collision. Thus, the end product of a collision between a uranium atom and a neutron can be a completely new and unnatural type of atom containing 93 protons, and which has been called neptunium.

Many of the 16 'transuranium' (beyond uranium) elements made so far have been created by this sort of neutron bombardment, each collision of a neutron generating a new proton for the target nucleus when the neutron splits into that proton and a fast moving electron. Some transuranium elements can be made in a slightly different way, by bombarding existing atoms with other particles such as the nuclei of helium or boron atoms, but the principle behind all such atom building feats is the same—an existing atom is bombarded with some particles which can, on colliding with the atom's nucleus, provide the new protons (and sometimes new neutrons) needed to change that nucleus into the nucleus of some other type of atom. The new atoms may initially carry an overall positive charge, of course, since they have acquired one or more positively charged protons, but they can soon collect electrons from chemicals in the environment to generate neutral atoms of a new and unnatural type. When we call these atoms unnatural, however, we really only mean that they are unnatural to the earth. Some or all of them may well exist naturally elsewhere in the universe.

In looking at how we can artificially make new atoms by combining existing nuclei with either neutrons or other nuclei, we are also seeing, in principle, how all of the natural atoms of the earth have been made by nature. The universe is believed to have begun as an expanding cloud of hydrogen, with all the other elements being made by the collision of hydrogen nuclei to produce helium nuclei, followed by further nuclear collisions which generated all types of atoms up to uranium. Most of these collisions, known as 'nuclear fusions', are believed to have occurred within the nuclear furnaces that we call stars, or during the spectacular 'supernovae' explosions which mark the death of many stars and the scatterring of their harvest of atoms out into space.

Protons, neutrons and electrons are the fundamental building blocks of atoms. These building blocks came together to form hydrogen, helium and small amounts of lithium some time after the big bang before there were any stars. As gravity pulled them together to light the fires of stars, these simplest atoms fused, and continue to fuse, to generate more complex atoms during the burning and final explosive deaths of the stars. Stars are the atom building machines which created the atomic diversity of the world. You and the world around you are made from the ashes of stars.

4   Energetic electrons

The search for knowledge tends to uncover new questions which would not initially have occurred to us almost as quickly as it reveals answers to the questions we originally wanted to answer. The knowledge summarized in the previous chapter leaves us feeling we know a bit more about what matter is, but it also presented some intriguing puzzles.

For example, it revealed that all atoms are composed of specific arrangements of protons, neutrons and electrons; it told us that the protons and neutrons are bound together into a tiny dense nucleus; and that the electrons are spread throughout large volumes of space around the nucleus by virtue of their possession of sufficient energy to defy the electric force which would otherwise pull them into the nucleus; but what is energy?

Chemistry is essentially concerned with two main things, electrons and energy, and the manner in which they interact. Electrons have been introduced to you and their two vital statistics—charge of -1, mass of about 0.000545 amu—revealed. Energy has been left hazy and undefined.

If you look to a science dictionary or encyclopaedia to find out what energy is, you will probably be told it is 'the capacity to do work' or 'the ability to do work'. So what is work? One common way of expressing the scientific concept of work is to say something like 'work is done when a weight is raised, or when a process occurs which could, in principle, be used to raise a weight.' The raising of a weight is simply serving here as a standard example of a process involving work, against which all other processes involving work can be measured. So if we say that a certain amount of work is done when a weight of 1 kilogram is raised by 1 metre, we would say an equivalent amount of work is done in any process which, if its work were somehow channelled into raising a weight, would result in a weight of 1 kilogram rising by 1 metre.

The idea of work being what is required to raise a weight probably fits quite easily with your everyday impression of the meaning of work. You have lifted weights, you know it is difficult, you know it is 'hard work'; but why is it difficult? It is difficult because it involves fighting against the force of gravity which attracts the weight and the ground towards one another. If you recall that gravity is one of the four fundamental forces of the universe, then you are on the verge of a deep revelation about what work and energy really are. The process of doing work, defined as raising a weight from the ground, is simply the process of moving something against the pull of the fundamental force of gravity. Gravity is used merely as the most familiar representative of the four fundamental forces. We could just as easily have defined work by saying 'work is done when an electron is pulled away from an atom's nucleus, or when a process occurs which could, in principle, be used to pull an electron away from an atom's nucleus.' You see, pulling a negatively charged electron away from a positively charged atomic nucleus is directly comparable to pulling a weight away from the surface of the earth. In one case we are struggling against the fundamental gravitational force, in the other against the fundamental electric force. So a more general definition of work would tell us that 'work is done when something is moved against the "will" of a fundamental force, or when a process occurs which, in principle, could be used to generate movement against a fundamental force.'

So what is energy, so far defined as just the capacity to do work? Clearly, energy must be the capacity to bring about movement against a fundamental force. The idea of energy becomes increasingly clear the more you think of it as some form of 'fundamental force defiance' or 'antiforce'.

Energy is not really a 'thing', it is more a 'state' associated with certain arrangements and motions of matter. Just as the stress that builds up within many human relationships is not a tangible thing, but more a state of affairs, so the possession of energy is more an intangible condition than a discrete and concrete thing. Really, energy is just an idea, an idea invented by the human mind to make sense of the world and the things that happen in it.

How can we measure and quantify the concept of energy? We can do it by defining some standard amount of energy, against which all other amounts can be compared. So, if energy is the ability to do work, and one standard measure of work is how high it will lift a weight, we can define a standard unit of energy by somehow quantifying what is happening when an object is lifted through a certain height, against the force of gravity.

The strength of a force is measured in units called 'Newtons', with 1 Newton defined as a force that will accelerate a 1 kilogram mass at a rate of 1 metre per second per second (i.e. every second it is under the influence of the force, the velocity of the mass will increase by 1 metre per second). A convenient standard unit of energy is therefore the 'Newton metre', which represents the energy it takes to move an object a distance of 1 metre *against* a force of 1 Newton. Another name for the Newton metre is the 'Joule', so energy is measured in Joules. If it takes 10 Joules to raise some weight by 10 metres, it will take 20 Joules to raise it by 20 metres. If it takes 10 Joules to raise the weight by 10 metres, it will take 20 Joules to raise twice the weight through 10 metres.

So the Joule is the standard unit of energy, against which we can measure and compare all other amounts of energy; and remember that when we talk about amounts of energy in terms of so many Joules, we are simply relating the energy to the height through which it would lift a certain weight if all the energy could somehow be used to power that task.

It is time we got back to the atom and its electrons, and the energy the electrons need to keep them orbiting around the nucleus rather than plummeting towards it.

Now that we have seen the way in which energy can be quantified, we can look at the 'energy levels' of the various orbitals available to electrons in an atom. Remember an orbital is just a region of space which can be occupied by up to two electrons, and in order to occupy any orbital, an electron must possess an appropriate amount of energy, sometimes called the energy level of the orbital, which is therefore the energy level of any electron occupying the orbital. The energies associated with different orbitals can be measured and presented in diagrams like figure 4.1.

This figure illustrates a fundamental aspect of the microworld: the energy content of that microworld varies discontinuously, or in other words it is 'quantized'. What that means is that entities in the microworld, such as electrons, seem to be distributed between a fixed series of distinct energy levels, rather than having energies spread freely over a continuous range. Of all the possible energies which electrons could have, they seem restricted to just a few energy levels as shown in the figure.

So, once again, we find a deficiency in the initially helpful analogy between electrons and satellites orbiting the earth. We can put a satellite in any orbit we like around the earth. We simply need to choose the amount of rocket fuel and the rocket trajectory needed to place our satellite anywhere between the lowest possible orbit and the highest possible one. Electrons seem much more restricted. The orbital space around an atom's nucleus is quantized into a series of specific orbitals corresponding to discrete defined energies of the electrons that occupy

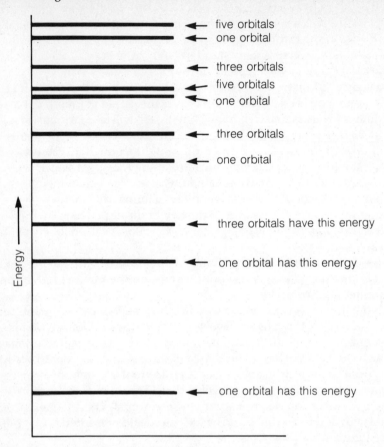

Figure 4.1   The orbitals available for the electrons of atoms are restricted to certain discrete 'quantized' energy levels.

them. So it seems electrons cannot be anywhere around the nucleus, but must slot into one of the predetermined orbitals. Even this seemingly refined description of the microworld's orbital order is a rather simplified view of a more complex deeper reality, but it is vastly more accurate than the simplistic view of electrons able to orbit at will around the nucleus.

An electron is not forever stuck in whatever orbital it happens to occupy. An electron in a high energy orbital can sometimes fall down into a lower energy one (for reasons explored later), an act which leaves us with another dilemma. Probably the most famous law of physics is the law of conservation of energy, which states that a reduction in energy in one place must be compensated for by an equivalent increase in energy somewhere else. It seems there is a fixed amount of energy in the universe, a quantity which can never increase or decrease overall. So the only possible energy transactions involve the *transfer* of energy from one place to another.

This ancient law had to be altered somewhat when Albert Einstein showed us that matter and energy are best regarded as two separate aspects of the same thing, which we could call 'mass–energy' or 'mattergy' if we wished. His famous equation $E = mc^2$ demonstrated that energy could be converted into matter, and matter into energy, something which happens all the time in the nuclear reactions that sustain the sun, or drive nuclear power plants or generate the destructive energy of a nuclear bomb from the calm solidity of uranium or plutonium. This complication is not a problem for us. In the first place, particles of matter are not created or destroyed during chemical reactions; and secondly, the first law of thermodynamics remains unbroken provided we adjust it to read 'the total amount of mass–energy in the universe is constant, with mass and energy representing two different manifestations of the one fundamental phenomenon.' So-called physical laws are merely man-made descriptions of the universe based on our observations. We are free to alter them in the light of new observations and experience. So, back to our dilemma: if an electron falls from a high energy orbital to a lower energy one, what happens to the excess energy? What happens, in other words, to the extra quantity of energy needed to keep the electron in the 'high' orbital, but not required, indeed not allowed, if it is to occupy the 'low' one? The answer is that the unneeded energy is released into the environment; and it is released in the form of 'electromagnetic radiation', a form of energy whose most familiar representative is light.

Light is energy on the move, and on the move very fast—speeding along at 300,000 kilometres per second, which is believed to be the fastest that anything can ever go. By 'energy on the move' we mean, of course, 'the ability to do some work on the move', so that when light interacts with some other matter, such as an electron, it can serve to push that matter up into a higher energy state.

So suppose the electron considered above gives out light energy corresponding to the energy difference between the high and low energy orbitals it falls between. That light energy could rush away at 300,000 kilometres per second, until it met another electron in another atom which happened to be in the low energy orbital. When this happens, the energy of the light could serve to 'kick up' the low energy electron into the high energy orbital, since the light has precisely the amount of energy required to do so—the light energy, in other words, could be 'absorbed' by the electron and serve to raise the electron's energy level. Overall, some energy will have been transferred from the electron in the first atom to the electron in the second atom, and it would have been transferred between the atoms in the form of light. A change will have taken place in the universe, a change powered by the transfer of energy from one place to another, but which leaves the total amount of energy in the universe the same as it was before the change took place. This is the first significant example of a change that we have looked at in any detail in the main text—at long last, you may be thinking, something has happened! Since chemistry is all about change, and this is our first specific example of a change, it is worth careful consideration. The essential feature of the change has been the transfer of energy. We will discover that in one way the universe is wonderfully simple, because the essential feature of every change in the universe is simply the transfer of energy. Our first change has shown us the principle behind all change. If we can only work out *why* energy should be transferred from place to place, we will have uncovered both the nature of and the reason for all the change that powers all of the chemistry, and all of the physics, and all of the biology, and all of the everything else, that happens in the universe we live in. We will explore that reason very soon, in chapter 5.

In the meantime, we need to examine energy a little more deeply, because there is another form of energy which has so far received only passing mention.

Up till now I have concentrated on a form of energy known as 'potential' energy. I have looked at the energy of weights raised from the ground and of electrons 'raised' away from the atomic nucleus. I have pointed out that both the weights and the electrons possess energy by virtue of their *positions*. These positions, high above the ground, or high above the nucleus, involve some defiance of a fundamental force, so they embody the clear potential for change, as the weight falls to the ground, driven by the gravitational force and accompanied by the release of some of its energy; or as the electron falls closer towards the nucleus, driven by the electric force and accompanied by the release of some of its energy. The energy an object possesses by virtue of its position, a position which defies some fundamental force, is known as potential energy simply because the situation has the clear potential to change if, for some reason, the force wins out and the object loses some energy as its defiance of the force is reduced or abolished.

Objects can also possess energy by virtue of their *movement* relative to other things, and this energy of movement or energy of motion is known as 'kinetic energy'. Potential energy and kinetic energy are the two fundamental and interconvertible forms of the phenomenon of energy. Many objects, such as the electrons in an atom, possess some of both.

Why do we say moving objects possess some energy as a result of their movement? Well, remember that we defined energy as the capacity to do work, and said that work is done when a weight is raised, or when a process occurs which could, in principle, be used to raise a weight. Look at figure 4.2 and you will see the motion of a rolling ball serving to raise the ball, i.e. raise the weight of the ball, as the movement slows and then stops. As the movement of the ball takes it up the slope, its motion decreases and then eventually ceases; but in return for the loss of the motion the weight of the ball has been raised to a certain height up the slope. A physicist would say that the kinetic energy of the ball has been converted into the potential energy stored in the higher position of the ball. What is really happening here is that the motion of the ball eventually begins to propel it into a position which requires some fundamental force to be defied—in this case the force of gravity. The motion of the ball 'wants' to take it further to the right, but in order to move further to the right it must raise the weight of the ball against the fundamental force of gravity.

This simple example illustrates the basic reason why moving objects possess an amount of energy that depends on their motion (strictly, it depends on how fast they are moving and how massive they are, according the the formula: kinetic energy $= \frac{1}{2}$ mass $\times$ velocity^2). The motion of moving objects can move them into positions which involve increasing defiance of a fundamental force, as a result of which the movement slows and eventually stops when all of the energy of movement (kinetic energy) is converted into energy of position (potential energy). That is the general principle of the interconversion of kinetic and potential energy; later on we will see it at work in situations more directly relevant to chemicals and chemical reactions.

Looking from figure 4.2(a) to 4.2(c), we see kinetic energy being converted into potential energy. Figure 4.2(d)–4.2(e) shows what would happen next, unless there were some sort of trap or ratchet mechanism to keep the weight at its new height and prevent it falling back down again. Without such a mechanism, the weight of the ball would cause it to start rolling down the slope again. The potential energy which was briefly stored in the height of the ball would be converted back into the kinetic energy of motion. This is essentially what happens when two objects collide and then bounce off one another—kinetic energy becomes potential energy, which is then converted back into kinetic energy. In the case of collisions, the potential energy is a result of the atoms of the colliding objects being forced too close together, in defiance of the electric force, rather than as a result

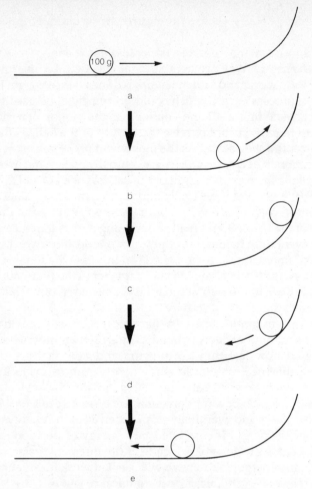

Figure 4.2 Kinetic energy becomes potential energy, becomes kinetic energy . . .

of a weight being raised against the gravitational force; but, as you will see more clearly below, what happens during all collisions also depends on the fundamental principle that potential and kinetic energy are two interconvertible forms of the same thing, or idea, which we call energy.

What, you may be asking, has all this got to do with atoms? It has a great deal to do with atoms, because all atoms are moving, as you were told in chapter 3, and all electrons can be considered as moving as well.

The chemical microworld is a seething sea of moving particles.

Consider one of the simplest things of the world, a sample of some element such as mercury, the fluid 'quicksilver' which so fascinated the ancients. A globule of mercury lying stationary on a laboratory bench contains only mercury atoms on the move. If we could peer into the silvery mass, right down to the level of atoms, we would see all the mercury atoms rushing about, bumping into one another and bouncing off one another. The globule is stationary because the motion of the atoms is chaotic and not in any preferred direction, so that it cancels out overall to produce no overall motion of the globule across the bench; but the globule is full of motion and kinetic energy nonetheless.

Chemical reactions are brought about by particles colliding with one another, but most collisions result in no chemical reaction at all, the particles merely 'bouncing' apart to dart off in different directions. The unproductive bouncing collision of two particles such as atoms is the simplest and most common interaction of the microworld. We should look at it more closely to try to understand why it happens.

When the motion of two atoms brings them together on a collision course, the electrons and nuclei of the approaching atoms begin to feel one another's presence in ways which cause both attractive and repulsive electric forces to build up. An attractive force develops between the nucleus of each atom and the electrons of the other atom; repulsive forces develop between the electrons of the converging atoms, and also between the nuclei of the converging atoms. All these forces, of course, develop simply due to the attraction between opposite electrical charges and the repulsion between like charges. When the two atoms are relatively far apart, the overall effect of these interatomic forces is an attractive one, in other words it tends to draw the atoms together; but when they actually collide a very strong overall repulsive force can begin to dominate. Essentially this repulsive force builds up because the negatively charged electrons of the two atoms are being pushed very close to one another, as are the positively charged nuclei. The collision, in other words, is forcing negative charge to approach negative charge and positive charge to approach positive charge in a way that is strongly opposed by the electric force. As a result, the movement of the atoms is dramatically slowed, then stopped, and then instantly reversed as the electric force pushes them violently apart. What I have just described, of course, is a 'head on' collision between two atoms. Most collisions are glancing blows but the principles remain the same— some of the kinetic energy of the atoms' movement being converted into potential energy on collision, and then back into kinetic energy as the atoms move apart.

So as this process of 'elastic collision' takes place, at least some of the initial kinetic energy of the approaching atoms is converted into the state of high potential energy associated with the two sets of electron orbitals being very close together and the two nuclei being very close together. In this state the shape of the electron orbitals becomes grossly distorted as they are repelled from one another. The potential energy is then converted back into the kinetic energy of the motion of the atoms as they bounce away from one another.

There is another way in which the strains imposed on the colliding atoms can be released, a way which involves the atoms reacting together rather than bouncing off one another (see chapter 6), but the process of elastic collision just described is essentially what happens whenever two particles (atoms, molecules or ions) collide and then bounce apart. The energy of the collision forces the colliding nuclei and electrons into highly strained orientations in which the electric force eventually restores order by pushing the colliding particles apart.

Whenever anything in the everyday world bounces off any other thing, the forces responsible are really the forces of electric repulsion which build up when the atoms (or other particles) of the objects are forced together against the will of the electric force. So the description of the collision of two atoms given above embodies the essence of the collision between all objects which come together and then bounce apart. The repulsive force between like electric charges explains why pool balls bounce off one another; why footballs bounce off feet when they are kicked, and off goalposts when they hit them; why bullets ricochet off walls, and so on . . . These repulsive electrical forces also explain why you cannot walk through walls, since an attempt to walk through a wall, or pass your hand through it, is really just a concerted collision between the particles of your body and the particles of the material of the wall. As you press your hand firmly against a wall, for example, the force you feel resisting you is really the force of electric repulsion between the sub-atomic particles within the atoms from which you and the wall are made—electrons of the wall repelling electrons of your hand, and protons of the wall repelling protons of your hand. This electric repulsion is also what prevents you from falling through the seat you are sitting on, or the floors and the ground you walk upon. So elastic collisions between particles such as atoms are very important and influential—the universe would be a very strange place without them.

Something of great significance happens when particles collide— their kinetic energy tends to be redistributed such that, if one initially had more energy than the other it passes some of that energy on to the other. Consider the extreme case of a moving pool ball striking a stationary one with a glancing blow. Obviously, after the collision both balls will be moving. The total kinetic energy of motion will be the same, but some of the moving ball's energy will have been transferred to the initially stationary ball. All this should fit very easily with everyday experience. If something moving fast collides with something moving slowly, the initially fast moving thing slows down while the initially slow moving thing speeds up. In collisions, in other words, the energy of motion tends to be dispersed from the faster moving objects to the slower moving ones. If you are walking along the street and a jogger bumps into you, the jogger will slow down and you will speed up!

So collisions are great redistributers of energy: they allow energy concentrated initially in one object to become dispersed throughout the motions of a greater number of objects. A fine example of that is the 'break' at the start of a pool game when the kinetic energy of the single fast moving cue ball becomes dispersed into and distributed among all the balls in the pack when the cue ball collides with the pack. A few pages ago I told you that the essential feature of all change was simply the transfer of energy, and that if only we could find out why energy should be transferred from place to place we would have uncovered both the essential nature of and the reason for all change. Well, we just have uncovered the reason why energy should be transferred, although we have sneaked up on it so stealthily that it may not yet be apparent to you. The energy transfers which make things happen occur because of the automatic and inevitable *dispersal* of energy from regions in which it is concentrated into regions in which there is less of it.

The essential nature of all change, including all chemical reactions, is the transfer of energy. The essential reason or guiding force behind that transfer is the inevitable dispersal of energy from regions of high energy to regions of lower energy. These are the simple facts at the heart of our universe and all its physical, chemical and biological complexity. In the next chapter I will begin to explain to you more clearly what they mean and why they are so important. The remainder of the book will illustrate the truth and power of this beautifully simple vision of our world.

Box 4.1 The wave-mechanical atom

The modern picture of atoms emerged from treating the movement of electrons as some sort of wave-like motion, rather than as the clearly defined orbital motion expected from discrete particles.

It has become clear that the true nature of the microworld

cannot be properly described using either the concept of particles or that of waves alone. Light, for example, was long held to consist of 'waves of energy' travelling through space, since, amongst other things, it can be reflected and refracted and diffracted just like everyday waves such as the waves on the surface of the sea; but in some ways light also behaves as if it were made up of a stream of discrete 'light particles' which physicists call photons. When electrons were first discovered some people thought that they were transmitted as waves much like light, but the prevailing view soon came to regard them as discrete particles, since that was how they often behaved. In the 1920s, however, scientists returned to the idea of electrons as waves and met with a startling success.

The German physicist Erwin Schrodinger took the idea of 'electron waves' to its mathematically logical conclusion. He took the known properties of electrons and incorporated them into a 'wave equation', which is simply a mathematical equation which allows the behaviour of any wave to be described in a quantitative manner (using numbers, in other words). Schrodinger's wave equation can yield an infinite number of solutions. In other words, it can describe an infinite range of possible electron-waves. If certain seemingly sensible restrictions are imposed upon the solutions we should accept, however, it becomes much more revealing and useful.

One particularly interesting type of wave is known as a 'standing wave' or 'stationary wave'. This is the sort of wave you get in a violin string, for example, with different parts of the string vibrating up and down while the wave is not moving along the string overall. Standing waves are essentially just vibrations which serve to store energy in whatever is vibrating, so they seem likely to be relevant to the energy stored within the electrons surrounding an atom. If only the standing waves generated by the Schrodinger equation are considered, then a series of standing waves around an atom emerges, whose energy matches the energy levels found for electrons within an atom. These standing waves are taken to represent the orbitals available for electrons around an atom's nucleus; and the Schrodinger equation also predicts the shape of all the electron orbitals corresponding to various energies (where the orbital 'shape' is really just the position of its '90 per cent probability boundary' see page 20).

So our picture of the electron orbitals available around atoms is derived from the standing waves of various energies and shapes predicted by the Schrodinger equation. These standing waves are waves in three dimensions, a bit like the vibrations within a metal sphere hit by a hammer, rather than the vibrations up and down of a violin string. They indicate that an electron around an atom behaves like a vibrating cloud of electrical charge, rather than a tiny hard particle rushing around the nucleus; and in every case the predictions from this 'wave-mechanical' view of the atom match very closely the actual activities of atoms which can be observed by experiment.

So what are electrons—particles or waves, solid objects or mysterious vibrations? The true answer is probably 'a bit of both but strictly speaking neither'. Remember that science constructs *models* of reality which allow us to analyse and predict the behaviour of reality. Some aspects of the behaviour of electrons are best described by treating them like waves, while other aspects are best described by treating them as particles, but that does not necessarily mean that they really are either waves or particles or some blend of both. We should not naively assume that the behaviour of the microworld must exactly match the behaviour of the things we call waves and particles up in the large-scale macro-world around us.

There is a 'wave–particle duality' found in all aspects of the microworld—light, protons, neutrons and all other particles and phenomena in addition to electrons. Working scientists rarely trouble themselves with this at first rather puzzling phenomenon. Instead they are just delighted that by treating the microworld as consisting of either waves, or particles, or a bit of both, whenever it suits them, they can glean knowledge which helps them to understand and predict that hidden world's behaviour. Treating electrons as waves has given us our most accurate picture so far of the structure of the atom, revealing all of the orbitals available for electrons around an atom—their shapes as well as their energies. That success is sufficient reason for treating electrons as waves whenever it suits us to do so.

The Schrodinger equation paints the following picture of the electron orbitals around atoms (see figure 4.3). It reveals an atom as surrounded by 'shells' and 'subshells' of orbitals, with room for more orbitals within each shell as we move farther from the nucleus.

The first shell has room for one orbital, and you have already been told that each orbital can contain up to two electrons, so there is room in the first shell for two electrons at most. Incidentally, for an orbital to contain two electrons there must be a

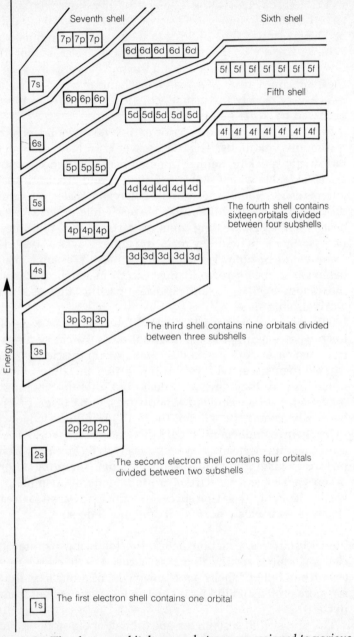

Figure 4. 3　The electron orbitals around atoms are assigned to various shells and subshells. (The subshells are known as 's', 'p', 'd' or 'f' subshells for rather technical historical reasons which need not concern us.)

rather subtle difference between the motion of these electrons. Chemists describe and visualize this required difference by imagining that electrons can spin, and for two electrons to occupy any one orbital they must be 'spinning' in opposite directions—clockwise and counterclockwise. This idea of opposite electron spins is useful, but it is just an analogy. It provides an easy way for us to visualize and think about a mathematical difference in the wave description of electrons which is more subtle than the spinning motions of balls in the everyday world.

The second electron shell contains a total of four orbitals, and so has room for eight electrons. You can see from figure 4.3, however, that this shell is split into two subshells of slightly different energies. The lowest energy subshell contains just one orbital, while the other three orbitals of this shell are of identical and slightly higher energies, which is why they are described together as a subshell. The crucial point about the orbitals of any subshell is that they all represent identical energy levels for the electrons within them.

The third electron shell contains nine orbitals, and so has room for 18 electrons. It is split into three subshells containing one, three and five orbitals respectively.

The fourth electron shell contains 16 orbitals, and so has room for 32 electrons. It is split into four subshells, containing one, three, five and seven orbitals respectively.

As we investigate the space around an atom's nucleus, seeking out the orbitals available for electrons at increasingly higher energies, we find the orbitals building up according to a clear pattern, and that pattern continues until we have orbitals available for all the 92 electrons of a uranium atom, which is the largest naturally occurring type of atom on earth; and can continue to provide orbitals for 'unearthly' man-made atoms such as neptunium (93 electrons), plutonium (94 electrons), and all the way to element number 109, with 109 electrons, which is as far as our atom-building efforts have got.

The essential feature of the electron orbital pattern is that as we move to higher energies, corresponding to orbitals whose outer boundaries are farther from the nucleus, we find room available for increasing numbers of orbitals and therefore increasing numbers of electrons.

When we examine any particular atom in the rather low energy everyday environment of the earth, we find another pleasing simplicity—electrons tend to occupy the lowest energy orbitals out of all the orbitals available. This state of an atom, in which all its electrons are in their lowest energy orbitals, is known as the atom's 'ground state'. There are many circumstances in which that state can be disturbed by the transfer of energy into the atom, in the form of electromagnetic radiation for example, to make electrons jump up into higher energy orbitals; but the ground state is our basic starting point for a consideration of atoms and their electrons.

The periodic table (figure 3.3) lists all the atoms of the world and allows us to work out how many electrons each contains (since the number of electrons equals the number of protons, which is the 'atomic number' of the atom). So the information in figure 3.3 and figure 4.3 allows us to identify the ground state electronic structure of an atom, simply by fitting the required number of electrons into the orbitals shown in figure 4.3, starting with the lowest energy orbital and working upwards until all the electrons are accommodated.* A hydrogen atom, for example, contains just one electron which will be found in the single orbital of the first electron shell—the lowest energy orbital of all.

A helium atom, with two electrons, will contain a filled first shell, rather than the half filled shell of a hydrogen atom.

A lithium atom contains three electrons, so while two of them can fit into the single orbital of the first shell, the third will be forced to occupy the lowest energy orbital of the next shell up.

A carbon atom contains six electrons, so its lowest energy ground state will have two electrons in the single orbital of the first shell, two in the single orbital of the lower energy subshell of the second shell, and the remaining two in two separate orbitals of the slightly higher energy subshell of the second shell. These two highest energy electrons occupy separate orbitals, rather than pairing off into one orbital, essentially because their negative electrical charge causes them to keep as far apart as possible.

We have now met all the main principles governing the electronic structure of atoms: electrons occupy the orbitals found

---

*In truth, that statement is a bit too bold since, sadly, there are a few exceptions to the neat pattern of atomic structure obtained by following the suggested procedure. For 20 of the 92 naturally occurring elements one or two electrons appear in different orbitals to the ones predicted. The rulebook of science is riddled with such exceptions. They do not invalidate the general use of the rulebook, but they do serve to remind us that our attempts to summarize reality in terms of very simple models are imperfect, however useful they are. There are good reasons for the exceptions to the general rules of atomic structure outlined above, but there is little point in considering them here.

around atomic nuclei whose energy and shape are predicted by Schrodinger's wave equation; they will normally occupy the lowest energy orbitals available; and if two or more electrons have orbitals of equal energy available to them they will occupy separate orbitals wherever possible.

So, to summarize how figure 4.3 allows us to work out the ground state electronic structure of an atom, we simply imagine ourselves filling up the available orbitals with the electrons we must fit in, always using low energy orbitals in preference to higher energy ones, and keeping electrons in separate orbitals whenever orbitals of identical energy are available.

If you have examined figure 4.3 closely, however, something may be troubling you. The figure clearly shows that the orbital referred to as the 4s orbital is of lower energy than the ones called the 3d orbitals. The 3d orbitals are clearly part of the third shell while the 4s one is part of the fourth shell, so why is part of the fourth shell filled before the third shell is complete? The answer you would glean from the table is the true one: the 4s orbital is filled first simply because, in an atom in which all orbitals up to the 3p ones are filled, the 4s orbital is a lower energy orbital than the 3d ones. That leaves the question of why are the 3d orbitals referred to as being in the third shell at all? The answer is that the number of shell which an orbital belongs to is a consequence of the mathematics of the wave equation. So the equation which tells us what orbitals are available indicates a mathematical link between all the orbitals referred to as, for example, 'third shell orbitals', and that is essentially why they are all called third shell orbitals, even though some of them (the 3d ones) are not filled with electrons until part of the fourth shell is filled. So there is a good mathematical reason for the shell structure outlined in figure 4.3, even though it introduces the complication and seeming contradiction that part of the fourth shell fills up before the third shell is completed. The numbering of the shells 'falls out' of the mathematics used to tell us what orbitals are available around atoms, it does not always accurately indicate the relative energies of the orbitals in different shells.

You will find similar complications higher up the figure if you care to look, such as the '5p' orbitals filling up before the '4f'. I hope I have persuaded you to accept that there is a good reason for such features of the figure, although I have not gone into the mathematics of the reason. We do not need to concern ourselves about such complexities when considering the electronic structure of any atom. The *energies* of the orbitals are what matter, rather than the labels we know them by, and the simple rule is that electrons will normally occupy lower energy orbitals in preference to higher energy ones. They will only jump up into higher energy ones if something makes them so jump, such as the arrival of energy in the form of electromagnetic radiation.

Box 4.2 The electromagnetic spectrum

Electromagnetic radiation, such as light, is 'energy on the move', and more specifically it is electromagnetic energy on the move. What does that really mean? It means that electromagnetic radiation is a travelling disturbance of the universe which has the ability to make things move against the will of the electric and magnetic forces, which are really two different aspects of the one electromagnetic force (see page 22). Electromagnetic energy behaves like a travelling vibration, or in other words like a wave, which can make electric charges accelerate or oscillate a bit like corks bobbing in response to the passage of water waves. Nobody knows for sure exactly what electromagnetic radiation is, and it is subject to the same puzzling 'wave–particle duality' as electrons. In addition to being described as a wave, in other words, it can also be regarded as consisting of discrete particles or packages of energy which are released in one location, to travel through space and be absorbed somewhere else. When electromagnetic radiation is released from something, of course, the energy of that something falls by an amount equal to the energy of the released radiation; and when the radiation is eventually absorbed by another thing, that other thing's energy increases by the same amount.

Light is the form of electromagnetic radiation we are most familiar with, because we can see it. There is, of course, a 'spectrum' of different colours of light, and these different colours of light differ in the 'frequency' of the electromagnetic waves, and, from the particle viewpoint, in the energy which each 'particle' (photon) of light possesses. The violet end of light's spectrum corresponds to waves of the highest frequency (i.e. the troughs and peaks of the travelling vibrations are closest together and each photon carries more energy than other forms of light); while the red end of the spectrum corresponds to waves of much lower frequency and lower energy.

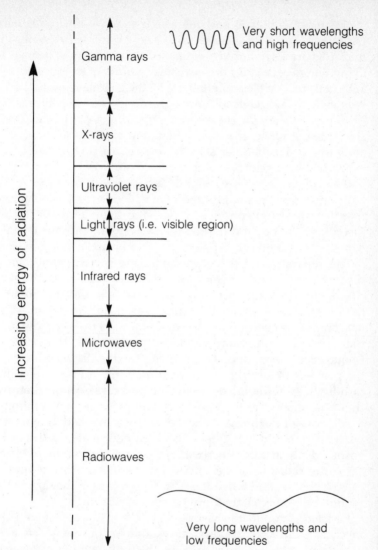

Figure 4.4   The electromagnetic spectrum.

This visible spectrum of light waves, however, represents only a tiny segment of the full spectrum of electromagnetic energy, since a wide range of invisible forms of electromagnetic radiation exists, corresponding to frequencies and energies both much higher and much lower than those associated with light (see figure 4.4).

The most energetic and highest frequency electromagnetic radiations are called gamma rays. These are released from the nuclei of various radioactive elements and can do great damage to living things because of the devastating input of energy they can subject us to. A bit less energetic than the gamma rays are the 'X-rays', which, although still very hazardous, can safely be used in low doses to examine the internal structure of our bodies. Next come the ultraviolet rays which have sufficient energy to give us sunburn, and sometimes skin cancer; and then we arrive at the very narrow range of frequencies which our eyes can respond to, and so which we can see. Slightly less energetic than the visible light rays there are the infrared rays which we feel as heat; and then the spectrum continues down into the microwave region (used for cooking) and the radio wave region whose waves we exploit to communicate with one another by radio and television.

The separation of the electromagnetic spectrum into these distinct wavebands is somewhat arbitrary. The spectrum really consists of a continuous range of radiations with all possible frequencies from the very high (more than a million trillion cycles of vibration per second as the waves pass) to the very low (less than a hundred thousand vibrations per second). This range of frequencies corresponds to wavelengths (the distance between the peaks of the waves of vibration) ranging from less than one hundredth of a billionth of a metre (gamma rays) to more than one thousand metres (radio waves), and, in principle, on to infinity.

All forms of electromagnetic radiation are released from matter when it settles down into a lower energy state, and can be absorbed into matter to increase its energy correspondingly. In all cases the radiation is generated by changes in the motion of electric charge, and when it is absorbed it causes changes in the motion of electric charge in whatever absorbs it.

Radiation around the visible region, and beyond it into the ultraviolet and X-ray regions, corresponds to the energy changes associated with electrons moving between atomic orbitals of different energy. Gamma radiation is associated with changes in the structure of atomic nuclei involving the rearrangement of their electric charge. Infrared radiation is released and absorbed when the chemical bonds linking atoms into molecules (which we will look at later) change their modes of vibrational stretching and bending and so on; while microwaves are released and absorbed when molecules change their speed of motion and rotation as they chaotically tumble around the microworld. Radio waves are most commonly released when electrons oscillate to and fro within an electrical conductor, the waves travelling through space to cause similar oscillations in other conductors such as receiving aerials.

In all cases, the fundamental effect is that changes in the motion of electric charges in one place bring about changes in the motion of electric charges elsewhere. The incredibly fast moving disturbances which link these changes are what we call electromagnetic radiation.

5 Dispersal drive

Each time you warm yourself by a fire, or by the heat of the sun, you are experiencing and exploiting the phenomenon which makes the universe work. You are experiencing the effects of the tendency of energy to disperse from places where it is concentrated into places where it is less abundant. This chapter explores the mechanisms behind that tendency in a very simple situation, to prepare you to recognize and appreciate the tendency whenever you meet it in the more complex phenomena of the chemical world.

Imagine you are holding a metal knife, which you rest on an ironing board with the bottom half of the knife touching a hot iron. You will be sensible enough to soon let go of the handle of the knife, since you know it will get hot; but why will it get hot, and what do we really mean by 'hot'? We can answer the last question very easily. The 'heat' of an object is related to how fast its particles are moving, so the atoms of a hot knife are moving faster than the atoms of a cold one. If you accidentally rest your hand on a hot surface, like the hotplate of an electric cooker, it is the movement of the atoms that jolts you into pain and a reflex withdrawal. If you step into a bath only to jump out gasping that it is 'too hot', you really mean that its water molecules are moving about too fast—a myriad invisible particles hammering into your tender skin and stimulating your sensitive nerves into action. The technical definition of heat says that it is a measure of the kinetic energy of the particles of an object, something which depends not only on the speed of the particles but also on their mass. In practical terms, however, the heat of any substance is really just a measure of how fast its particles are moving, since their masses remain constant but, as the substance heats or cools, the particles simply move more quickly or more slowly. When we use a thermometer to observe the way in which something's temperature rises or falls, we are gathering information about the speed of its particles and the way in which that speed is changing.

So that is what we mean when we say that our knife brought into contact with a hot iron will heat up; but why will it heat up? Clearly, the metal surface of the iron is hot because its atoms are moving about very fast and so with a lot of kinetic energy. When we allow our knife to come into contact with the hot iron, the knife's slow moving 'cool' metal atoms come into contact with the fast moving 'hot' metal atoms of the iron; and 'by come into contact' I mean that collisions begin to occur between the iron's fast moving atoms and the slower moving atoms of our knife. We have already seen what happens when moving objects collide. Their combined kinetic energy tends to be redistributed more evenly, meaning that an object with lots of kinetic energy loses some of that energy when it bumps into a less energetic object, while the object with less kinetic energy gains more kinetic energy as a result of the collision. So the atoms at the end of your knife, the end that is in direct contact with the heat of the iron, are going to speed up simply because they are colliding with many fast moving energetic atoms of the hot iron. The end of the knife, in other words, gets hot, while it brings about a corresponding cooling of the part of the iron from which its heat is drawn. Heat energy is automatically dispersing from a hot region into a cooler one, simply because 'it has no other choice'. It has no choice because, in a world of moving particles participating in random collisions, the energy of motion is inevitably dispersed throughout all of the particles towards an even distribution.

This is such an important principle that we should contemplate it a little further, even at the risk of labouring the point. The actual movement of the atoms, both in the knife and in the metal surface of the iron, is chaotic. All the atoms in the knife are not moving at the same speed, but travel at a range of speeds corresponding to a range of kinetic energies, and the same applies to the atoms of the iron; but *on average*, the atoms of the iron are moving with much greater kinetic energy than those of the knife—that is really all we mean when we say that the metal of the iron is hotter than that of the knife. This does not mean that there are no atoms in the knife moving with more energy than the atoms of the iron they collide with—there may be a few; but the number of such collisions, which will disperse energy away from the knife and into the iron, will be very much smaller than the number of collisions taking energy the other way. The *probability* of any individual collision dispersing energy into the knife is much higher than the probability that it will disperse energy out of it, simply because most of the atoms of the knife will be moving about less energetically than the atoms of the iron they collide with.

The dispersal of energy towards a more even distribution is the guiding force of all change in the universe, but it occurs as an inevitable consequence of the chaotic movement of the objects which carry that energy, rather than in some smooth, ordered and ideal manner. It happens simply because, when all the motions and other activities of objects are considered, there are more ways in which energy can be dispersed than ways in which it can become concentrated. In the case we are considering, there are more ways for energy to be dispersed into the knife than ways for it to travel in the other direction simply because there are inevitably going to be more collisions between atoms that will transfer energy in that direction, for the reasons considered above. The automatic and inevitable dispersal of heat energy does not stop at the end of the knife. The hot fast moving atoms at the end of the knife will jostle into the cooler slower moving ones beside them, and so the heat will travel up the knife towards the end you were holding in your hand. If you have been sensible enough to let go in good time the heat will begin to disperse out into the air and into the ironing board due to collisions between the energetic atoms of the knife and the slow moving particles of the air and ironing board. If you have not let go, the jostling atoms of the knife will collide with particles in your skin, which will speed up and in turn jostle the heat deep into the tissues of your hand. You will cry out and let go of the knife, perhaps shaking your head at the pain that can be caused by an encounter with the laws of thermodynamics!

The specific law which would be causing you pain is the 'second law of thermodynamics', which is arguably both the simplest and the most significant physical law of all. It can be stated in many forms, the easiest of which is 'energy disperses.' More formally, 'In any change, energy will always move, overall, towards a more even distribution.' The 'overall' is crucially important, as we shall see later.

Another important consequence of the dispersal of energy, a consequence, in other words, of the second law, is that it allows chemicals to change between the 'states' of solid, liquid and gas. Imagine that we have placed a lump of metal, such as lead, in some very sturdy receptacle, and have put it right at the heart of a furnace. As the heat of the furnace is jostled into the lead the lead gets hotter, until eventually it turns to liquid. What has happened is that the lead atoms have begun to move around so quickly that they can easily slip around and over one another. The 'solid' structure of the lead has collapsed and we are left with a hot molten liquid. Eventually, as the heating continues, some of the lead atoms near the surface of the liquid begin to move so fast that they can escape from the liquid and travel off alone through the air around them. They have left the liquid and become gas. If we then remove the receptacle from the furnace, the remaining liquid lead will slowly cool as its heat disperses into surroundings which are now cooler than itself. Eventually the lead will revert into the state of cold solid lead; and if we gathered enough gaseous lead and allowed it to cool, we would see liquid and then solid lead reforming as well.

Solid, liquid and gas are the three clearly distinguishable 'states of matter', and substances change between these states when they are heated or cooled. The particles of solids are close together and jostling around relatively slowly, the particles of liquids are a little farther apart and moving a bit more quickly, the particles of a gas are very far apart and moving extremely fast.

So we have examined some of the consequences of the movement of particles of matter and the collisions which that movement brings about, and that has allowed us to identify the fundamental principle described by the second law of thermodynamics—the principle that in any change energy will always move, overall, towards a more even distribution. The direction of change in the universe is governed by this 'dispersal drive', by which I mean that all change must move in the direction that promotes the dispersal of energy, overall, towards a more even distribution. That is why 'hot' things cool down and why 'cool' things can warm up. It is why the sun can warm you and a dip in the sea can cool and refresh you. It is why ice cubes melt in your drinks, but do not suddenly form and grow within a hot cup of tea. Dispersal drive guides all change in the universe, and it does so because in a world of chaotically moving particles there are always more ways for energy to become more dispersed, overall, throughout the particles, than for it to become concentrated in small groups of particles. Natural change is guided by the competing probabilities of energy dispersal and energy concentration, with dispersal always winning overall because the 'game' is inevitably biased in its favour.

The technical term for this fundamental and inevitable rule of the universe is the 'increase in entropy' of the universe. Entropy is a difficult concept to explain in detail, and it can be defined in several different ways. If you study chemistry to a higher level than this simple introduction, you will eventually meet these various definitions, but regardless of how it is defined the central simplicity of the idea of entropy remains the same. Whenever scientists talk about an increase in entropy, they are talking about the dispersal of some of the universe's energy towards a more even distribution, no matter how that process is quantified and defined. If you keep that central simplicity in mind, any further and deeper exploration of chemistry will be made much easier and more enlightening.

We have not yet encountered any real chemistry in this discussion of the direction of natural change. Changes in a substance's temperature, and the transitions between its solid, liquid and gaseous states are all 'physical changes' in which the chemical character of the substance remains essentially the same. Solid lead and liquid lead and lead gas are all composed of lead atoms, just as solid iron and liquid iron and iron gas are all composed of iron atoms; in no case does any chemical change occur.

Chemistry begins when particles do not bounce off of one another when they collide, but instead 'react' to yield different chemicals from the ones that took part in the collisions. The overall direction of chemical change is governed by energy dispersal just like the overall direction of the transfer of heat, but often in more subtle and less obvious ways. Chapter 6 explores this alternative consequence of collision, revealing the principles behind all of the chemical reactions which lie at the heart of all chemistry.

6   Binding force

I was twelve years old and I stood before the class with a burning taper in my hand. Clamped to a retort stand on the bench beside me ballooned a large plastic bag filled with oxygen and hydrogen gas. Our teacher had filled it carefully a few moments before—roughly two parts hydrogen for one part oxygen—then he had selected me as the poor fool who was to set it alight. He tried to make his lessons entertaining, but to frightened youngsters his ideas of entertainment often seemed more akin to sadism. He ordered the rest of the class to crouch behind the solid wooden bench at the far side of the room, with only their eyes and the tops of their heads visible, and then as he commanded me to hold the taper at arm's length and let its flame play up through the narrow open neck of the bag he too ducked behind a bench and shouted at me to look away.

I looked away, and I trembled with fear, but I knew I would have to get the thing done. I closed my eyes while my outstretched hand moved closer to the bag. Eventually, the flame must have found the gas. There was an astonishing explosion and for an instant I felt I was about to die as its thunder cracked against my ears and its blast blew through my hair. Then there was silence, and I was still alive; and as my classmates began to cheer I knew what it felt like to be a hero.

As I stood recovering from my ordeal I think I appreciated for the first time the true power of chemistry. The teacher picked up a few shreds of polythene and held them up before the class. He ran his finger across the plastic and we saw droplets of water dribbling across it and down his hand. He announced that this was all that remained of the great volume of gas which I had ignited—the oxygen and the hydrogen had reacted together to form water. Two chemicals had combined to create another quite different chemical. A chemical reaction had taken place.

When chemicals 'react' what they are reacting to is the collisions between the particles they are composed of. Two or more particles collide and react by giving rise to something new. The simplest reaction

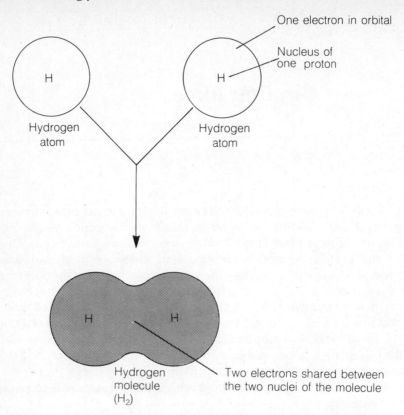

Figure 6.1   The formation of a hydrogen molecule.

will involve the simplest atoms, which you should remember are hydrogen atoms (consisting of one proton and one orbiting electron). When hydrogen atoms collide they often simply bounce off one another, but sometimes they react to form a hydrogen 'molecule'. Figure 6.1 shows you how. A hydrogen molecule is merely a re-arranged form of two hydrogen atoms, in which the identity of the individual atoms is lost as they merge into a new single particle. In this new particle the electrons are shared between the two protons (in other words between the two nuclei) rather than each orbiting around just one of them. In a sense the hydrogen molecule can be regarded as an atom with two nuclei.

Whenever atoms combine by sharing electrons we say that a molecule has been formed. A molecule can be defined as a particle in which two or more atoms are combined by virtue of sharing some electrons between them. We describe the combination by saying that the original atoms are held together by a 'chemical bond'. Why should chemical bonds such as the one shown in figure 6.1 be created? What makes this chemical reaction happen?

Chemical reactions are what happen when chemicals collide and the electromagnetic force pushes and pulls their electrons and nuclei into new arrangements. So the kinetic energy of motion and the electromagnetic force are the initiating agents of chemical change, although our consideration of the second law leads us to look for the dispersal of energy as the guiding force controlling the direction of that change.

Imagine two hydrogen atoms approaching one another. As they approach, new forces of electric attraction and repulsion come into play. We have already considered the possible effects of the repulsive forces which build up between the two atoms' electrons and nuclei— these are the forces which cause atoms to bounce off one another—but forces of attraction must be considered too. As the atoms approach, the electron in each atom becomes attracted to the other atom's nucleus as well as to its own; and similarly, each nucleus becomes attracted to the other atom's electron as well as its own. When the collisions between the atoms occur with energies and orientations which allow the repulsive forces to dominate the atoms bounce apart; but some collisions bring the atoms together in ways that allow the attractive forces to win because they initiate a complete rearrangement of the electrons which, overall, results in less resistance against all the various pushings and pullings of the electromagnetic force. Such a new arrangement, in other words, is a lower energy arrangement in which the 'will' of the electromagnetic force is more satisfied than in the free atomic arrangement. The lower energy arrangement is the one in which the two electrons are shared between the two nuclei, leaving the atoms joined together by a chemical bond.

But what happens to the energy left over as a result of the atoms settling into their new lower energy molecular arrangement? The first law of thermodynamics tells us it cannot disappear, but must be transferred somewhere else. Some of it may be lost in the form of electromagnetic radiation, given out as electrons fall down from high energy orbitals occupied during the chaos of collision to the lower energy orbitals of the final molecular state. Many chemical reactions give out light, such as the light of an explosion, as electrons fall down from high energy orbitals which they are temporarily pushed up into during the reaction collision. Energy lost in the electronic rearrangement of a reaction can also be stored for a while within the products of the reaction, such as a hydrogen molecule, in the form of internal vibrations of the new molecule. Chemical bonds are a bit like springs, and the two atoms of the hydrogen molecule can bounce in and out as the bond stretches and compresses in turn. As it stretches, it moves the molecule into a higher energy arrangement because it is stretching against the pull of the attractive forces holding the molecule together, and as it compresses it moves the molecule into another high energy arrangement because it is compressing the nuclei too close to one another against the force of electric repulsion between them. So a bond that is vibrating in and out stores some energy, just as a spring vibrating in and out holds more energy than one sitting quietly in its unstretched and uncompressed state. This energy of vibration is soon lost, however, as it becomes dispersed out to the surroundings as a result of the collisions between our new hydrogen molecule and the other particles all around it.

Just as a fast moving molecule slows and a slow moving one speeds up when the two collide, so a fast vibrating bond will lose some of its energy of vibration as its molecule collides with other molecules which are moving more slowly and whose bonds are vibrating less. Collisions between particles cause all forms of motion to become dispersed towards an even distribution: not just 'straight line motion', but also all the vibrations and flexions of bonds and any spinning motions of a molecule overall or some of its parts.

Molecules are a bit like tangled balls of springs, with each spring representing a bond. The balls of springs move about and rotate and collide, and as they do so both their overall kinetic energies and the energy of all their vibrations and flexions and rotations tend to become evenly distributed throughout the entire 'balls of springs' population.

So the energy difference between the free atomic state of two hydrogen atoms, and the combined molecular state, is soon dispersed into the gross and the internal motions of all the particles all around. This can leave the new molecule trapped in its new lower energy molecular form—trapped because the energy which could raise it back into the free atomic form has dispersed away into the surroundings. A chemical reaction has taken place because the motion of two atoms brought them together and the electromagnetic force pulled their electrons and nuclei into a new arrangement, but the change has been fixed or stabilized by the dispersal away of the energy lost in the process. The reaction has been given direction and permanence by the dispersal of energy—the fundamental guiding force of all change.

When my chemistry teacher mixed together hydrogen and oxygen gas to generate the mini bomb for me to ignite, he was mixing together hydrogen molecules and oxygen molecules, rather than single atoms. At everyday temperatures and pressures hydrogen comes in the form of H_2 molecules and oxygen comes in the form of O_2 molecules, because in both cases the free atoms tend to react together to form the lower energy molecular form. Although hydrogen and oxygen tend to exist in the form of molecules rather than free atoms, they are still elements, since a sample of hydrogen contains nothing but hydrogen atoms, albeit combined into molecules; and a sample of oxygen contains nothing but oxygen atoms, combined into molecules. When hydrogen and oxygen react with each other, however, they can form a 'chemical compound' (water) in which atoms of different elements, hydrogen and oxygen, are chemically bonded together.

By far the majority of chemical substances are compounds rather than elements. There are 92 elements available on earth, but an almost limitless variety of compounds formed by combining atoms of these elements. The compound we call water, created in the explosive reaction which so alarmed me in my youth, provides a good and simple example.

As I stood trembling before the class the molecules of hydrogen and oxygen in their polythene bag were rushing around and bumping into one another, but at this stage they were simply colliding and then bouncing off rather than reacting. They were colliding and bouncing off because they did not have sufficient energy to react. The reaction between hydrogen and oxygen to form water involves a major rearrangement of electrons, a rearrangement which proceeds by initially raising the internal energy of the molecules before it can be allowed to fall down into the new and lower energy arrangement of water. This means that for the reaction to take place electrons and nuclei must initially be pushed into configurations which strongly resist or defy the electromagnetic force, something which will not happen unless the collisions between the hydrogen and oxygen molecules are violent enough to bring about a considerable disruption of the balanced structure of the molecules.

When I gathered my courage and allowed the flame to play up into the mixture of gas in the bag its heat provided the energy needed to speed up a few of the molecules of hydrogen and oxygen which the bag contained. Just a little speeding up in a tiny portion of the gas mixture was enough. It jolted a few molecules into colliding with such force that the existing electron arrangements were totally disrupted allowing new arrangements to be explored. The new arrangement which the hydrogen and oxygen atoms adopted is shown in figure 6.2. Each oxygen atom ends up sharing electrons with two hydrogen atoms. The total positive electric charge of the three nuclei is $8+1+1$ (i.e. $+10$), while the total number of electrons in the molecule is also $8+1+1$ (giving a negative charge of -10). The water molecule is electrically neutral overall, and its electrons and nuclei are in a considerably lower energy arrangement than they were in when part of either oxygen or hydrogen molecules.

As the first few water molecules began to form, the energy released by their formation served to speed up some neighbouring hydrogen and oxygen molecules, causing them to react and release the energy needed to jolt more hydrogen and oxygen into reacting, and so on . . . The reaction proceeded as a self-fueling process which soon released energy so fast and in such large amounts that up in our slow moving macroworld we perceived it as an almost instantaneous explosion which shattered our ears and sent my hair flying in its blast.

Exactly the same explosive reaction powers the main engines of the American space shuttle as it blasts out of the earth's atmosphere and into orbit. The large tank secured to the shuttle's belly carries liquified oxygen and hydrogen which are combined and ignited to generate an awesome explosive force. It is a force which has been used successfully to hurl many shuttle missions into orbit (assisted by the solid fuel booster rockets). The force of many chemical reactions, however, while useful if tamed and carefully controlled, can be devastating if allowed to proceed out of control. The devastating force of the reaction between hydrogen and oxygen was revealed to us all in January 1986, when we saw it destroy the shuttle Challenger. The power of chemistry can be harnessed to achieve the otherwise impossible, but if it runs out of control it can destroy us in instantaneous explosion and conflagration, or by slower and more subtle means.

We should now return to the water molecules which stream from the back of an ascending shuttle or which dripped from the shreds of polythene left over from my own initiation into the power of this reaction. The three constituent atoms of a water molecule are held together by the sharing of electrons between them, as are the two hydrogen atoms of a hydrogen molecule or the two oxygen atoms of an oxygen molecule. All bonds which hold atoms together by virtue of the sharing of electrons are known as 'covalent bonds', but there is an important difference between the covalent bonds holding each of the O_2 and H_2 molecules together and the ones holding the H_2O molecule together. In an H_2 molecule the electrons are shared equally between the two hydrogen nuclei, since both nuclei carry the same amount of positive charge: $+1$. It is, after all, the positively charged nuclei which attract the negatively charged electrons and so share them between one another. By 'shared equally' we mean that the electron orbital containing the shared electrons is perfectly symmetrical in both its shape and the extent to which its different parts are occupied by electrons, rather than being biased towards any one nucleus. Another way of thinking about it is to say that each electron effectively spends

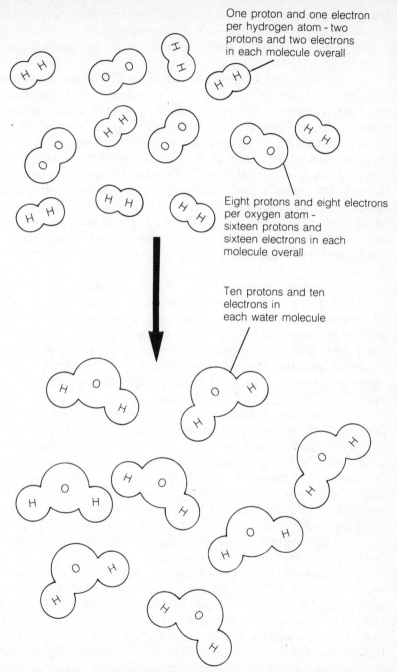

One proton and one electron
per hydrogen atom - two
protons and two electrons
in each molecule overall

Eight protons and eight electrons
per oxygen atom -
sixteen protons and
sixteen electrons in each
molecule overall

Ten protons and ten
electrons in
each water molecule

Figure 6.2   Hydrogen molecules and oxygen molecules combine to form water molecules.

the same amount of time close to one nucleus as to the other. A similar situation exists in the O_2 molecule: the two oxygen nuclei are identical, each carrying a +8 charge, so the electrons are shared equally between them.

In a water molecule, however, the electrons which hold the molecule together are shared between an oxygen nucleus with a charge of +8 and two hydrogen nuclei each with a charge of +1. Obviously, the electrons are more strongly attracted towards the oxygen nucleus than to the hydrogen nuclei, so they 'spend more time' around the oxygen nucleus; or, in other words, the orbitals occupied by the shared electrons are not symmetrical, but are distorted in favour of the oxygen nucleus. This means that the molecule carries a slight negative charge in the region of the oxygen nucleus, since that region has a greater share of the shared electrons, while there is a slight positive charge in the region of the hydrogen atoms. Such 'slight' charges are denoted as δ^+ and δ^- respectively, to indicate that they are not full positive or negative charges.

Covalent bonds formed by the *unequal* sharing of electrons are described as 'polar covalent' bonds, since the unequal electron distribution causes a 'polarization' of electric charge into regions of slight positive (δ^+) and slight negative (δ^-) charge. Obviously the extent of this polarization depends on the difference between the electron-attracting power—the 'electronegativity'—of the specific nuclei involved (see box 6.1). Some polar covalent bonds are more polarized than others.

We have seen one of the two main ways in which chemicals can react: they can react to form new structures in which electrons are *shared* (equally or, more commonly, unequally) between the atoms involved in new ways. When two chemicals come together and combine due to the sharing of electrons, covalent bonds have been formed (one bond per pair of shared electrons see box 6.2). There is one other major way in which chemicals can combine. They can combine as a result of the complete *transfer* of electrons from one chemical to another. The simple example of the common 'table salt' we sprinkle on our food will serve to illustrate this type of chemical bond.

Consider a sodium atom and a chlorine atom, shown in very simple form in figure 6.3. When sodium and chlorine react together each sodium atom 'donates' an electron to a chlorine atom, as shown in the figure. This creates a new situation for us. The sodium atom has been converted into a particle which is no longer electrically neutral. It has a nuclear charge of +11 now counterbalanced by only 10 electrons rather than the original 11. So the overall charge of the particle will now be +1, and instead of a sodium atom it is described as a sodium 'ion' with a

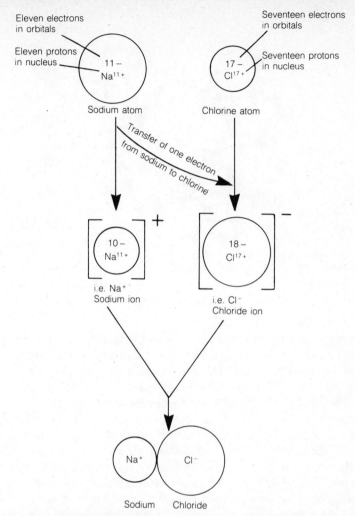

Figure 6.3 The reaction of sodium and chlorine atoms to form sodium chloride, held together by an ionic bond. (Chlorine normally exists as Cl_2 molecules but a free atom has been used here for simplicity. When sodium reacts with Cl_2 molecules each sodium atom donates an electron to one chlorine atom overall, so the principles and end results of the reaction are the same.)

charge of $+1$ (denoted Na^+). An ion is a particle with an overall electrical charge. The original chlorine atom has also been converted into an ion (called a chloride ion), but in this case the charge on the ion is negative since by gaining an extra electron it carries one electron more than is needed to neutralize the charge on the nucleus. The Na^+ and Cl^- ions

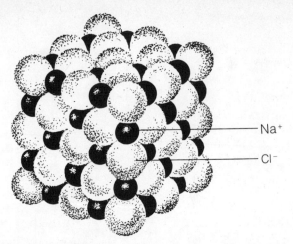

Figure 6.4   A tiny portion of the crystal lattice of sodium chloride. This simple arrangement of sodium and chloride ions extends in all dimensions to create visible crystals.

are strongly attracted towards one another by the electromagnetic force, since they carry charges of opposite sign. So as soon as they are formed the ions move together and become 'stuck' to one another by what is known as an 'ionic bond'. This bond is essentially just the force of attraction between the positive and negative ions.

If you could peer down to the level of the ions in a grain of salt you would see a situation similar to figure 6.4: a vast three dimensional array of sodium and chloride ions, containing one sodium ion for every chloride ion and all held together by ionic bonds to form what is known as an ionic 'lattice' structure. There are no molecules in this lattice, since molecules are electrically neutral particles composed of two or more atoms held together by covalent bonds, and there are no free atoms, since all atoms are electrically neutral overall. Instead there are just ions—the third and final basic type of particle involved in chemistry. The particles of chemistry are atoms, molecules and ions, which are themselves composed of the sub-atomic particles: protons, neutrons and electrons.

Having been introduced to an ionic bond you should be wondering why it forms. Why do sodium atoms donate electrons to chlorine atoms and why do the chlorine atoms accept them? By now you should expect the answer to involve energy and its dispersal. The final state of figure 6.3, with a sodium and a chloride ion nestling against one another, is a lower energy configuration than that at the start of the reaction, in which the sodium and chlorine atoms are free and unreacted. The energy released by the formation of this ionically bonded state will normally be dispersed away into the surroundings, leaving the sodium and the chlorine trapped in their new ionic form. So reactions which generate ions and ionic bonds occur for the same reasons as reactions which generate molecules and covalent bonds—they occur because the inevitable tendency of energy to disperse makes them do so.

The ions of many ionic compounds carry multiple charges, such as $+2$, -2, $+3$ or -3, as a result of losing or gaining more than one electron during their formation (for reasons outlined in box 7.2). Such ions assemble into crystal lattices held together by ionic bonds just like the sodium and chloride ions considered above, although the precise structures of the lattices of different ionic compounds are different. If the charges on the positive and negative ions are of equal magnitude, such as $+2$ and -2, then the ions will again occupy their lattice in a simple 1:1 proportion. If the magnitude of the charges on the positive and negative ions differs, then the ions will be present in the proportions needed to create an electrically neutral lattice overall. So in a lattice formed from ions of charges $+2$ and -1, we will find two of the negative ions for every one positive ion, and so on for other situations of unbalanced ionic charge.

One further complexity is that some ions are so-called 'molecular' or 'compound' ions, consisting of several atoms covalently bonded together but carrying a positive or negative electrical charge overall. Examples are ammonium ions(NH_4^+), consisting of a nitrogen atom covalently bonded to four hydrogen atoms with the entire complex carrying a charge of $+1$; and carbonate ions (CO_3^{2-}), consisting of a carbon atom covalently bonded to three oxygen atoms and all with a charge of -2 overall. Such complex ions also assemble into crystal lattices, each lattice containing ions in the proportions required to make an electrically neutral structure overall, with all the ions held together by the ionic bonds between them.

You have now met the major types of strong bonds which hold matter together to give it its impression of substance and solidity. These are the covalent, polar covalent and ionic bonds discussed in this chapter, along with the 'metallic bonds' responsible for holding metals together and discussed in chapter 3. In all cases, of course, the same 'binding force' is responsible for bonding different atoms and ions into larger assemblies: namely the electric force which draws electric charges of opposite sign towards one another and pushes charges of like sign apart.

What we have not yet considered in any detail are the forces which exist *between* distinct chemicals before and after they react, and even if they do not react together at all. These are weaker and more subtle forces than the true chemical bonds we have considered so far, but they are both vital and very powerful controllers of chemistry nonetheless. We will discover that these too all stem from the effects of the electric force which holds such dominance over the world of chemistry.

I have described that chemical world as one in which there are two basic types of bonds capable of binding atoms or ions together into chemical compounds. These are the covalent bonds generated by the sharing of electrons between atoms, and the ionic bonds holding together positive and negative ions created by the transfer of electrons from one chemical species to another. In addition to the 'full' charges associated with ions, many chemicals also carry the partial charges associated with polar covalent bonds. Many chemicals, in other words, carry the regions of slight positive charge (denoted δ^+), found at one end of a polar covalent bond, and the regions of slight negative charge (denoted δ^-) found at the other end of polar covalent bonds. Such regions of partial charge on different molecules can serve to draw the molecules together and hold them in a loose complex by the power of electric attraction.

Water molecules provide an excellent example (see figure 6.5). Remember that the hydrogen atoms of water molecules carry δ^+ partial charges while the oxygen atoms carry a δ^- partial charge (because the O–H bond is a polar covalent one). As you can see from figure 6.5, this allows the electric force to pull water molecules into a loosely bonded network in which all of the hydrogen atoms are held close to oxygen atoms, and vice versa. There is a force of attraction, in other words, between neighbouring water molecules, thanks to their regions of δ^+ and δ^- partial charge. This attractive force is very weak compared to proper bonds, and so the cage-like structure of figure 6.5 is easily disrupted by the input of some heat energy. It extends throughout large volumes of water only when that water is frozen into ice, but even in a glass of water at room temperature there will be small regions of this cage-like structure, and all of the water molecules, even though many may be tumbling over and around one another, experience a weak attractive force between them due to the attraction between their δ^+ and δ^- regions and the corresponding oppositely charged regions of other molecules. This attraction has some very significant effects. It makes the boiling temperature of water, for example, 100 degrees centigrade. Without its effect water would boil at a much lower temperature, since less heat energy would be needed to separate the molecules of the liquid into the free individual molecules found in a gas. So without these weak interactions between water molecules life as we know it could never have arisen on earth. All of the water within the

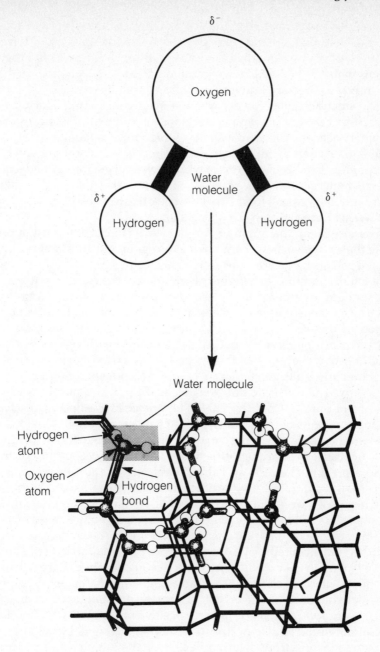

Figure 6.5   Hydrogen bonding in water (see text for details).

cells of our bodies, as well as the water of the seas and rivers and lakes, would have boiled away into gas. So these intermolecular forces, though weak, are nonetheless very significant. Whenever hydrogen atoms with a δ^+ charge play a central role in such weak bonds (as they do in water) they are known as 'hydrogen bonds'.

The intermolecular forces we have been considering, due to the attraction between δ^+ and δ^- regions of different molecules, are nowhere near as strong as covalent or ionic bonds, yet they are still very influential. There is an even weaker force of attraction between all atoms and all molecules, yet one which has a very significant effect on many chemicals and their reactions. These weakest of all intermolecular (and interatomic) forces are known, after their discoverer, as Van der Waals forces or Van der Waals bonds.

To understand the origin of Van der Waals forces you must appreciate that on the surface of any chemical, regions of very slight positive and negative charge are constantly being formed and then destroyed, due to the continually changing distribution of the electrons. It is as if the random orbiting of all the electrons continually creates and then destroys regions where electrons have become momentarily concentrated in one region and momentarily depleted elsewhere. So the surface of all chemicals is covered in transient 'flickerings' of positive and negative charge, and this creates an electromagnetic attraction which gently pulls all chemicals towards all others, assuming other more powerful forces do not over rule it. This attraction occurs when regions of transient positive charge on one chemical find themselves opposite regions of transient negative charge on another. Obviously, the two chemicals will be drawn together; and in being drawn together they will cause their mutual attraction to be reinforced since the positive charge on one chemical will draw electrons of the other chemical towards it, and so reinforce the other chemical's negative charge, while that negative charge will repel the electrons of the first chemical and so reinforce its positive charge. Of course it is just as likely that regions of the *same* charge on different chemicals will find themselves next to one another, but such repulsive interactions will not reinforce themselves, but will tend to destroy themselves. A fleeting negative charge on one chemical will repel electrons from any neighbouring region of fleeting negative charge on another chemical, and so encourage that charge to disappear. Two regions of fleeting positive charge on neighbouring chemicals would tend to destroy one another in a similar way, by each attracting the electrons which will destroy the other charge.

So there are very weak Van der Waals forces operating between all neighbouring molecules and atoms which will tend to draw the particles gently together, unless they are swamped by other more powerful effects. Once again, although each individual region of attraction exerts a very weak effect, the overall effect of these weak interactions, averaged out across the surface of large molecules, for example, can be very significant indeed.

So the electromagnetic force is the binding force of chemistry: binding electrons and protons together into atoms, binding atoms together into molecules, binding ions together into the crystal lattices of ionic compounds, and causing the particles of chemistry to be gently attracted towards one another in various subtle ways. It is not only the force that binds, however, but also the force that pushes chemicals apart. Its competing powers of attraction and repulsion lie behind all of the manoeuvrings of matter which we call chemical change.

Box 6.1 Electronegativity

Chemistry can be viewed as a contest between nuclei for the electrons which they are so strongly attracted to. As chemical reactions proceed, the nuclei of chemicals 'fight it out' to gain the strongest possible hold on all the electrons involved. In some cases atoms end up sharing electrons equally in covalent bonds; in others they share the electrons unequally in polar covalent bonds; while sometimes electrons are completely won by some atoms, making them negative ions, and completely lost by other atoms which become positive ions.

How well any atom is likely to fare in the contest for electrons depends on how strongly its nucleus can pull electrons towards it. That, in turn, depends on the size of the nuclear charge, and the extent to which that charge is 'screened' by the electrons already surrounding it. The 'electronegativity' of an atom is a quantitative measure of how strongly it can pull electrons towards itself. The most electronegative atoms are found towards the top right of the periodic table (figure 3.3), the least electronegative towards the bottom left; and in general the farther apart any two atoms are in the table the greater will be the difference in their electronegativities. These trends mean that elements which are very far apart in the table tend to react to form ionic compounds, since one element will be much more electronegative than the other and so atoms of that element will pull electrons off the other type of atom. Elements which are a bit closer together in the table tend to form compounds in which polar covalent bonds dominate, with the degree of polarization being less the closer together and therefore more matched in electronegativity that the elements are.

Electronegativity is one of the central concepts of chemistry. It indicates each atom's 'ranking' in the grand competition for electrons which lies at the heart of all chemical change. It allows us to make sense of the results of each individual contest for electrons when we observe chemical reactions, and allows us to predict the likely results in reactions we have not yet observed.

Box 6.2 Theories of bonding

In the main text you were told that atoms can combine to form chemical compounds in which the constituent atoms are held together by chemical bonds. You were told of two principal types of bonds: covalent bonds, in which some electrons are shared between atoms (shared equally in pure covalent bonds and unequally in polar covalent bonds); and ionic bonds, in which one or more electrons are transferred from one atom to another to create positive and negative ions which then become held together in an ionic bond. In this closer look at bonding the first thing to realize is that pure covalent and pure ionic bonding are best regarded as the two extremes of a graded continuum of bond types. If we regard all bonds as situations in which electrons are redistributed around two or more nuclei, instead of just one nucleus, then the perfectly equal sharing of covalent bonding and the totally unequal sharing of ionic bonding represent the two extremes of redistribution. Most bonds lie somewhere in between. We tend to refer to bonds as either covalent (including polar covalent) or ionic, depending on how close to one or other of the extremes they lie, but that is really just a rather sloppy simplification which usually works fine but can sometimes land us in trouble.

In the box on 'The wave-mechanical atom' (box 4.1), you were told how electrons can be treated as waves and described very successfully using the Schrodinger wave equation; so what does this wave-mechanical view of electrons have to say about chemical bonding? It turns out that the Schrodinger equation can describe the behaviour of electrons in chemical bonds very well.

The mathematics can become very complex, forcing various simplifying assumptions on us unless we have powerful computers and lots of time available, but in essence the electrons of chemical compounds appear to behave in a wave-like manner just as they do in the more simple situation of free atoms. We must look at some simple examples to grasp the principles.

We have seen how a hydrogen molecule can be formed when two hydrogen atoms approach and share their two electrons between them. One way of using the wave equation to describe this reaction is to take the mathematical descriptions of the two separate atomic orbitals and *combine them*—a relatively simple mathematical operation. Doing this actually yields two results, in other words it yields a mathematical description of two new 'molecular orbitals', meaning orbitals in which the electrons surround both hydrogen nuclei rather than only one. It is a general principle of the mathematics involved that if you combine two atomic orbitals you get two possible molecular orbitals as a result, and if you combine four atomic orbitals you get four possible molecular orbitals as a result, and so on. So the total number of possible orbitals remains unchanged during the procedure. If we combine the two occupied atomic orbitals of two hydrogen atoms:

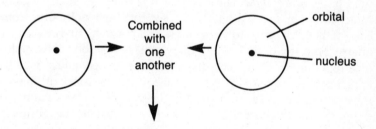

we get two molecular orbitals which look like this:

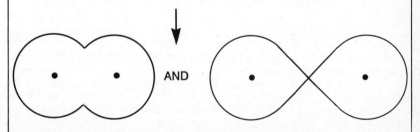

The first of these orbitals is a much lower energy orbital than the other. It is lower energy because there is a significant concentration of electrons in the region *between* the two positively charged nuclei, resulting in a reasonably balanced mix of regions of positive and negative charge (which are, of course, attracted to one another).

The other orbital has the negative charge concentrated at either end of the whole structure, which is a much higher energy arrangement since it leaves the two positively charged nuclei exposed to one another without much negative charge between them.

The first molecular orbital is known as a 'bonding orbital', because if the two electrons occupy it they will bond the two nuclei together by virtue of the attraction of the nuclei for the negative charge between them.

The other molecular orbital is called an 'anti-bonding' one, because if the two electrons occupy it then repulsion between the bared nuclei will tend to force them apart and encourage the break up of the complex back towards free atoms.

Now remember the principles that guided us when we looked at the filling of atomic orbitals by electrons. Electrons will normally occupy low energy orbitals in preference to higher energy ones, with a maxiumum of two electrons being allowed within any one orbital. The same rules apply to the occupancy of molecular orbitals. We have two electrons available, one from each of the hydrogen atoms, so these will normally fit into the low energy bonding orbital, where they will serve to bond the molecule together. That, essentially, is the wave-mechanical way of explaining the formation of chemical bonds: the atomic orbitals of the reacting atoms are combined to generate an eqivalent number of molecular orbitals, half of which will usually be bonding orbitals and half anti-bonding.* The electrons then occupy the lowest energy orbitals available, and if that leads to more bonding orbitals being occupied than anti-bonding ones then the molecule will be held together by one or more covalent bonds.

*A few molecular orbitals turn out to be what are known as 'non-bonding orbitals'. These have the same energy as the atomic orbitals which combine to form them, and they neither encourage nor discourage the bonding together of the atoms concerned. In a simple treatment of bonding, therefore, they can be ignored, since it is the imbalance between the occupancy by electrons of bonding and anti-bonding orbitals which determines whether or not a molecule is a feasible stable structure.

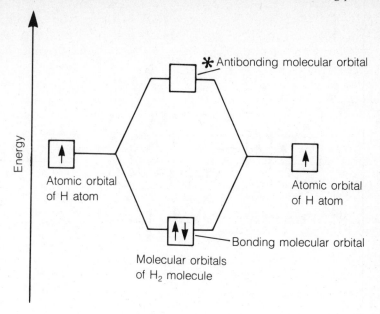

Figure 6. 6   Atomic and molecular orbital energies for hydrogen atoms and molecules.

By definition, if two atoms are held together due to the effect of one excess filled bonding orbital, i.e. two bonding electrons, they are held by one covalent bond. If they are held together due to the effect of two excess filled bonding orbitals, they are held by two covalent bonds, and so on. So two electrons occupying a bonding orbital consitute a covalent bond, provided their bonding effect is not cancelled out by a filled anti-bonding orbital.

The previous paragraph introduced some complications and subtleties which we can only fully appreciate by looking at some slightly more complex molecules. A new type of energy diagram will help us to do so. Figure 6.6 shows the energy levels of the atomic and molecular orbitals of hydrogen atoms and molecules which were discussed above. It shows that the two atomic orbitals combine to produce a bonding molecular orbital of lower energy than the atomic orbitals, and an anti-bonding molecular orbital of higher energy. The two available electrons can both fit into the lower energy bonding orbital (provided they spin in opposite directions, as indicated by the 'up' and 'down' arrows). The final

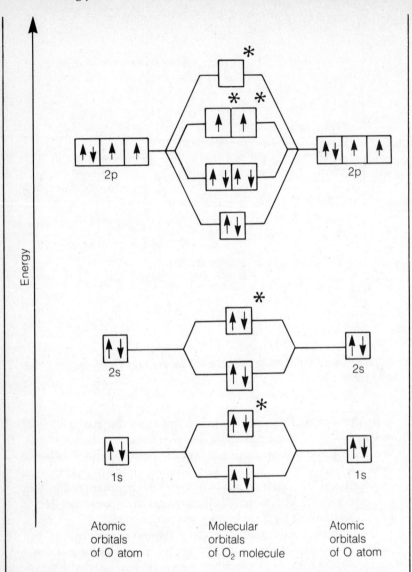

Figure 6.7   Atomic and molecular orbital energies for oxygen atoms and molecules.

configuration of the hydrogen molecule has two electrons sitting in a bonding orbital, and therefore bonding the two hydrogen atoms together by a covalent bond.

Two oxygen atoms can also combine to form a molecule

composed of two atoms, but since oxygen atoms contain eight electrons the situation is rather more complex. Figure 6.7 shows you the orbitals occupied by electrons in oxygen atoms, as well as the molecular orbitals created by combining the respective orbitals of two oxygen atoms. There are complexities evident in this figure which I do not want to go into, but the essential principles are quite clear and easy to understand. Electrons are found in five separate orbitals on each oxygen atom (referring back to figure 4.3 will allow you to confirm this, given that each oxygen atom contains eight electrons), and these combine to yield a total of ten molecular orbitals, of which five are bonding orbitals and five are anti-bonding (anti-bonding orbitals always being labelled with an asterisk). The 16 electrons available to the oxygen molecule then fit into the available orbitals, always occupying the lowest energy orbitals possible, to yield the pattern of orbital filling indicated in the centre of the figure. If you examine the figure closely you will find a total of 10 electrons in bonding orbitals and six in anti-bonding orbitals. So there is an excess of two pairs of electrons in bonding orbitals. You have been told that one pair of excess electrons in a bonding orbital constitutes one covalent bond, so the oxygen molecule is clearly bound together by two covalent bonds, which can be indicated by writing the molecule's structure as $O=O$, rather than $O—O$. The essential point is that when the effects of all the bonding and anti-bonding electrons are added together, there are four more bonding elec-trons than anti-bonding ones, and so the two oxygen atoms are bonded together by two covalent bonds.

Two nitrogen atoms can also combine to form a nitrogen molecule (N_2), and figure 6.8 reveals how. Each nitrogen atom has seven electrons, which originally occupy five orbitals. So once again there are 10 molecular orbitals available, but this time when all 14 electrons have been fitted in, 10 of them are found in bonding orbitals and only four in anti-bonding orbitals. So there is an excess of six (i.e. three pairs of) bonding electrons over the anti-bonding ones, corresponding to three covalent bonds. So a nitrogen molecule can be represented as $N≡N$, indicating the three covalent bonds that hold its two atoms together.

In all of our examples of bonding so far, we have considered every single electron on all of the atoms taking part. This becomes extremely difficult and complicated as we move on to larger, more complex atoms and compounds, and it is usually quite unneces-sary. It turns out that the deep 'inner core' of electrons in large atoms can largely be ignored when we consider these atoms'

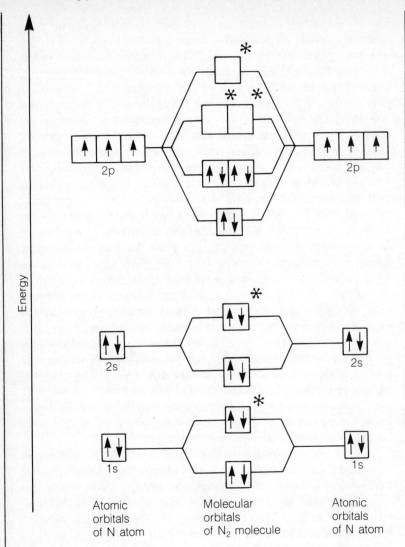

Figure 6.8   Atomic and molecular orbital energies for nitrogen atoms and molecules.

reactions, since the outer electrons are the only ones that have a significant net effect on the reactions and bonds that these atoms participate in. Look at figures 6.7 and 6.8, for example, and imagine that we ignored the bottom two molecular orbitals, which are the orbitals formed by combining the inner 'first shell' electrons of the atoms involved. The bonding and anti-bonding electrons in these orbitals exactly cancel one another out, so we could quite easily have disregarded them. So it is common for chemists to ignore such inner electrons if they know that their effects are going to cancel out.

Actually, other much more radical simplifying tricks are often performed by chemists when they try to describe and visualize the bonds between atoms, especially the bonds between atoms in large multi-atom molecules. These 'tricks' involve using a seemingly quite different and in many ways simpler model or theory of chemical bonding. Remember that I told you in chapter 2 that chemists sometimes use different models of chemical reality to describe the one real situation. Chemical bonding theory offers one of the best examples of this, with two main models available and each one providing a reasonably accurate description of what is really going on.

The model which most chemists believe to be the most rigorous, or to correspond most closely to true reality, in other words, is the full 'molecular orbital' model outlined above. In this model we envisage all of the atomic orbitals of all of the atoms of a molecule being combined to produce molecular orbitals which spread across the entirety of the molecule. So in this most rigorous approach the molecule is essentially regarded as an atom with many nuclei, and every electron has a certain probability of being found anywhere throughout the molecule, although there are, of course, regions of very high electron probability and regions of very low probability, just as there are electrons which will almost always be found around some specific nuclei rather than any others. The simplified version of this model considers molecular orbitals made from only the outer shell electron orbitals of the participating atoms, with the more tightly bound inner electrons of every atom being assumed to remain bound entirely to the original atoms, and therefore ignored. This simpler molecular orbital model again yields very satisfactory descriptions of the electron arrangements and therefore chemical activities of most molecules.

The more radically different alternative, however, is a model known as the 'valence bond' approach to bonding. I can outline the principles of this model by looking at the very simple compound, water.

Water is formed by the combination of one oxygen atom with two hydrogen atoms, as already shown in figure 6.2. Now in *principle* we could calculate the precise orbital motion of all 10 electrons involved, eight from the oxygen atom and one from each of the hydrogen atoms, by constructing molecular orbitals covering the entire molecule, or alternatively we could forget about the inner shell electrons of the oxygen atom (hydrogen atoms have only one electron and therefore only one shell), to develop a more simplified molecular orbital approach. In practice, however, a very satisfactory understanding of the situation is gained by considering only the two electrons of the hydrogen atoms, and two of the outermost electrons of the oxygen atom, and by imagining the atomic orbitals containing these electrons combining into bonding orbitals which connect only neighbouring atoms (see figure 6.9). This is the valence bond approach, so called because it only considers the outer or 'valence' electrons of the atoms concerned, and it envisages molecules being held together by restricted bonds localized between only two atoms.

Two of the outer electrons of the oxygen atom occupy half-filled orbitals, or, in other words, they occupy these orbitals on their own. If the orbital occupied by one of these outer oxygen electrons is imagined to overlap and then combine with the orbital of the electron of one of the hydrogen atoms, a localized bonding orbital is obtained which surrounds only the oxygen atom and that one hydrogen atom. The two electrons, one from the oxygen and one from the hydrogen, can occupy this bonding orbital together, leaving the oxygen and hydrogen atoms held together by a single covalent bond. Doing the same thing for the electron of the other hydrogen atom and another electron from the oxygen atom yields another covalent bond holding the other hydrogen to the oxygen. Notice that in this model a covalent bond corresponds to a bonding orbital localized around only two atoms and occupied by two electrons.

This alternative view of bonding yields very similar descriptions of molecules and predictions of their structure, reactivity and so on, as does the probably more accurate molecular orbital model. So chemists often use the valence bond model to simplify their consideration of more complex molecules, allowing them to restrict themselves to the consideration of bonding orbitals connecting only neighbouring atoms, rather than trying to conceive of the molecular orbitals covering the entire molecule. Some molecules, such as molecules of DNA (see page 131), can contain many thousands if not millions of atoms! Rigorously to calculate all of the true molecular orbitals encompassing the entirety of such massive molecules is simply not feasible. It is also unnecessary, because the simpler approach of constructing local-

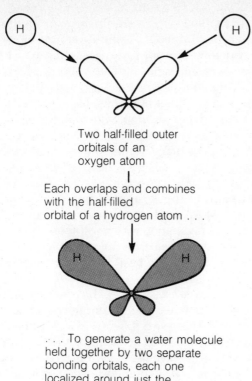

Two half-filled outer
orbitals of an
oxygen atom

Each overlaps and combines
with the half-filled
orbital of a hydrogen atom . . .

. . . To generate a water molecule
held together by two separate
bonding orbitals, each one
localized around just the
oxygen nucleus and the
nucleus of one of the
hydrogens,

i.e.

localized polar covalent bonds

*Figure 6.9 The formation of a water molecule according to valence bond
theory.*

ized orbitals surrounding only neighbouring atoms works
perfectly well, for most purposes. It yields a description of the
chemicals concerned which allows their behaviour to be both
predicted and explained, and that is all that chemists are trying to
do when they create their models of chemicals and chemical

processes. The more rigorous and more strictly accurate approach
of constructing molecular orbitals which are spread over all the
atoms of a molecule can be performed for relatively simple molecules using computers, but this is not normally necessary, and it is
not the way in which most chemists usually think about what is
happening when bonds are being formed.

The valence bond approach is older than the molecular orbital
approach, easier to visualize when dealing with complex molecules, and more accurately summarizes the way in which most of
the older chemists of the world actually think about the chemicals
they work with. Many younger chemists feel much more at ease
with the molecular orbital approach, so we can expect it to become
increasingly favoured as new textbooks appear, but it would be
wrong to suggest that the valence bond approach is merely an old-
fashioned and inaccurate approximation which will soon die out.
The debate over which method is the best is not yet settled, with
each approach seeming to provide the best match with experi-
mental reality in different situations.

What we are looking at here is a classic case of how science can
proceed quite successfully with two different and in some ways
competing models of reality. These two different bonding theories offer us two distinct ways of looking at chemical bonds, both
of which have their merits. They have yet to be replaced by one
unifying theory which will, one hopes, combine the best elements
of both models to give us an even better model of what is really
happening down in the unseen chemical microworld.

So, in summary, when atoms react to become joined together
by covalent bonds, we rationalize that by envisaging the combination of certain electron orbitals of the atoms to produce new
orbitals surrounding two or more nuclei rather than just one.
These new orbitals can come in bonding and anti-bonding forms,
and electrons normally occupy the lowest energy orbitals available. Each pair of electrons in a bonding orbital between two
atoms, in excess of the number of pairs of anti-bonding electrons,
constitutes a covalent bond. In most cases, however, this will not
be a pure covalent bond in which the electrons are shared equally
between the two atoms, but will be a polar covalent bond in which
one of the atoms gains a larger share of the electrons due to the
greater pull of its nucleus on these electrons.

If this glimpse into some of the complexities of chemical
bonding has confused you, then you should simply hang on to
the central simplicities:

atoms become covalently bonded together by sharing some

or all of their electrons between them, so that these electrons are 'in orbit' around more than one atomic nucleus; with equal sharing yielding pure covalent bonds and unequal sharing yielding polar covalent bonds. Ionic bonds are formed when electrons are transferred from one chemical to another, leaving one as a negatively charged ion and the other as a positively charged ion, and causing these two ions to be drawn together into an ionic bond. Pure covalent and pure ionic bonding are rare extremes of a graded continuum; most bonds display some characteristics of both these extremes, although either the covalency or ionicity of any specific bond tends to be its dominant characteristic.

7 Towards equilibrium

The platitude that a little knowledge is a dangerous thing is so often true because it can be misleading to understand only part of a complex situation. The knowledge which I hope you have gleaned from this book so far may have given you a very misleading impression of chemistry. It may have led you to believe that all chemical reactions proceed by changing chemicals into *lower energy* products, and that all chemical reactions proceed exclusively in one particular direction. Both these views are very common, but they are mistaken. They are impressions which you might understandably have gained from our considerations so far of the way in which the tendency of energy to disperse makes chemical change happen; but we must now correct them by looking more closely at the more subtle aspects of chemical behaviour.

Return to our fundamental 'touchstone' concept that all change in the universe is guided by the inevitable tendency of energy to disperse, overall, towards a more even distribution; and return also to the very simple physical system of a piece of hot iron. Suppose we have pulled this piece of iron from a furnace and thrown it into a vat of water to cool. It cools because most of the iron atoms are moving with greater energy than the water molecules with which they are constantly colliding, so there is a greater opportunity for collisions to take place which move heat energy out of the iron than into it.

Note that I say a *greater* opportunity, implying that the reverse process is feasible on rare occasions. At first, when the iron is very hot, there will be so little chance of a water molecule bumping into a less energetic iron atom that we can really discount such collisions altogether. At first there is a virtual one-way flood of heat energy out of the iron and into the water. But as the iron cools and the water heats up the average kinetic energies of the particles in both substances move closer together, and remember we must consider *average* energies, since the speeds and hence the kinetic energies of the individual particles will vary throughout some range. Eventually the ranges will begin to

84

overlap to a significant extent. In other words, some of the fastest moving water molecules will be moving with greater kinetic energy than some of the slowest moving iron atoms. Now we will see heat beginning to be transferred back into the iron whenever one of the fastest moving water molecules collides with one of the slowest moving iron atoms. If we measured the temperature of the iron and the water, we would still find the iron was hotter, and we would still observe the iron to be cooling and the water to be heating up, since the vast majority of collisions would still involve iron atoms moving with more energy than the water molecules they collide with. But this apparent one-way transfer of heat is really the overall or 'net' effect of a two-way transfer process in which the transfers in different directions are proceeding at different rates. Heat is moving out of the iron, when energetic iron atoms collide with less energetic water molecules, and it is also moving into the iron, when fast-moving energetic water molecules collide with some of the slower-moving, less energetic iron atoms; but since the first type of collision is more common, heat is moving out of the iron and into the water overall.

Obviously, as the iron cools further and the water heats up, the difference between the rates of outward and inward heat transfers will decline. The probability of water molecules colliding with iron atoms moving with less kinetic energy than themselves will increase (because the average speed of the water molecules is increasing while the average speed of the iron atoms is decreasing), while the probability of iron atoms colliding with less energetic water molecules will decrease.

The difference between the rates at which the two types of collision are occurring will decline until, eventually, it becomes zero and the two processes are proceeding to equal extents. What is the situation then? To anyone following events it would appear that all change had ceased. They would say that the iron has cooled, the water heated up, and now that both are at the same temperature the change brought about by dropping the iron into the water would appear to be over. (I am ignoring the complication that the contents of the vat will slowly pass their heat into the surroundings, since to consider that would only involve an extension of the same principles discussed here without yielding any further enlightenment. We will assume the vat is so well insulated that we can ignore this complication.)

Despite the apparent calm, however, all change will not have ceased. Down in the chemical microworld of atoms and molecules there will still be much frantic activity. Water molecules and iron atoms will still be bumping into one another, causing heat to pass into the iron whenever the water molecule has most kinetic energy, and to be transfered out of the iron whenever the iron atom has most kinetic energy. So change is still occurring, energy is still being dispersed in every individual collision, but no change and no energy dispersal is occurring overall because two equal and opposite processes are occurring at the same rates.

The simple system of iron and water has reached a position known as a dynamic equilibrium. It is an equilibrium, i.e. a 'balance' because nothing is changing overall; but it is a dynamic rather than a static equilibrium, because it is sustained by two processes which are constantly proceeding in opposite directions to equal extents. If we regard the transfer of heat from iron into water as the 'forward' process (simply because it was initially the dominant one) and the transfer of heat from water into iron as the reverse process, we can see that a state of dynamic equilibrium is reached when the forward and reverse processes are equally likely, and therefore proceed at equal rates.

This simple situation has profound significance to chemical reactions, for it is an analogy for what happens in all reactions to a greater or lesser extent. Every chemical reaction is, in principle, reversible. The extent to which the reverse reaction actually proceeds depends simply on how likely or 'probable' it is (just as the rate of heat transfer back into the iron depended on the probability of the required collisions taking place); and the probability of a 'reverse' reaction, just like the probability of the forward reaction, is determined by the inevitable tendency of energy to disperse, overall, towards an even distribution. So, enough of hot metal and water and simple analogies, let us look at a real chemical reaction.

The stability of Western society is dependent on the manufacture of ammonia, a chemical compound consisting of one nitrogen atom covalently bonded to three hydrogen atoms, and therefore having the chemical 'formula' NH_3 (see figure 7.1). We need ammonia to make fertilizers. It contains nitrogen in a form which can be used by plants directly, or which can be converted into other types of nitrogen-containing fertilizers. Without our supplies of ammonia and the fertilizers it yields the efficiency of our farms would dramatically decline. The economic 'problem' of surplus food would be replaced by a food supply crisis. We might adapt to less intensive and more natural forms of 'organic' farming eventually, but for a while, at least, our economies would be thrown into disarray and much hunger or even mass starvation would result.

Ammonia is manufactured by a chemical reaction between nitrogen (N_2) and hydrogen (H_2), the hydrogen being obtained largely (in the UK) from the 'natural gas' called methane (CH_4), while the nitrogen is readily available from air (air is 80 per cent nitrogen).

We can summarize what happens *overall* during the manufacture of ammonia, by writing a 'chemical equation'.

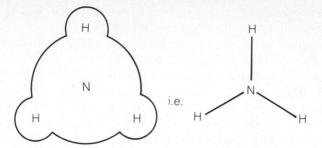

Figure 7.1   Ammonia.

A chemical equation lists the chemical formulae (i.e. the symbolic representations of chemicals which we have seen used already) of all the starting materials (reactants) and products of a reaction. So to create the equation for the formation of ammonia we need to write

$$N_2 + H_2 \rightarrow NH_3$$

It is a fundamental principle of chemistry, however, that atoms cannot be created or destroyed during a chemical reaction, merely *rearranged*. So chemicals must react in the proportions that will allow every atom on one side of an equation to appear somewhere on the other side. The equation written above implies that one nitrogen atom disappears during the reaction while an extra hydrogen atom miraculously appears—both quite impossible. The true situation is represented by the modified equation below

$$N_2 + 3H_2 \rightarrow 2NH_3$$

which is simply a 'shorthand' way of saying that for every one molecule of nitrogen consumed in the reaction, three molecules of hydrogen are consumed, yielding two molecules of ammonia; or in other words, one molecule of nitrogen reacts with three molecules of hydrogen to yield two molecules of ammonia.

The equation is now 'balanced', meaning that there are equal numbers of each individual type of atom on each side of the equation— two nitrogen atoms on each side and six hydrogen atoms on each side. Atoms have not been created or destroyed, just rearranged. Chemical reactions are mere rearrangements of matter.

Chemical equations summarize what happens *overall* during the course of a reaction, they do not tell us anything about the precise route or 'mechanism' of the reaction. A naive view of the equation for the formation of ammonia would suggest that a nitrogen molecule and three hydrogen molecules must simultaneously collide to generate two ammonia molecules out of the chaos of the collision. This is not what happens, although nobody knows for sure exactly what does happen. What is certain is that the simultaneous collision of a nitrogen molecule and three hydrogen molecules is far too unlikely for it to be the major route by which this reaction proceeds. Chemical reactions tend to proceed via many small steps producing high energy 'intermediates' which survive for such a short time they are very difficult to study. Chemists could write down various quite plausible mechanisms by which nitrogen and hydrogen could combine to form ammonia, but they still do not know which ones are the correct, or at least the most common, ones. The first step in the reaction is probably for a nitrogen molecule to collide with just one hydrogen molecule, generating a short lived and high energy intermediate which will react further when it collides with another hydrogen molecule, and so on. The details do not really matter, so we should return to the principles.

The formation of ammonia is actually a very reluctant reaction in normal circumstances. By reluctant, I simply mean that it proceeds very slowly. Nevertheless it does proceed, so let us assume that we have all the days, weeks, years or even decades that we need to watch a mixture of nitrogen and hydrogen react; and to begin with we shall assume that we have the mixture of gases sealed within a thoroughly insulated vessel, so that heat can neither escape from it nor enter into it. Ideally we would arrange for the temperature and pressure of the gas mixture to be rather high within the vessel to ensure that enough of the nitrogen and hydrogen molecules have sufficient energy to react when they collide, and to ensure that they collide quite often; but we can forget about all such considerations for the moment since I have already decided that the events within the vessel will be so interesting we can take all the time in the world to watch them unfold!

As the nitrogen and hydrogen molecules speed around within the vessel they will bump into one another and sometimes initiate the reaction which yields ammonia molecules. This reaction is one of those which leaves the products (two ammonia molecules) with less energy than the reactants (one nitrogen and three hydrogen molecules). The excess energy given out by the reaction disperses into the surroundings as heat. In other words it serves to increase the average speed of all the particles in the sealed vessel.

As soon as some ammonia molecules are formed, the reverse reaction, $2NH_3 \rightarrow N_2 + 3H_2$, becomes a possibility, although at first it is rather unlikely to proceed. It is unlikely to proceed because it requires ammonia molecules to collide with enough force to 'push them up' into the higher energy state of N_2 and H_2 molecules. Notice that if this does happen it is driven by the dispersal of energy *into* the chemicals concerned, rather than out of them, so it will only proceed to any great extent when energy has as much opportunity to disperse into ammonia molecules to form nitrogen and hydrogen as it has to disperse out of nitrogen and hydrogen to form ammonia.

If we think carefully about what is happening as the reaction proceeds, we shall see that all the time factors are operating which make the forward reaction less likely and the reverse reaction more likely. The forward reaction becomes less likely because the starting materials, nitrogen and hydrogen, are becoming used up, causing the probability of the collisions which initiate the reaction continuously to decrease. At the same time the number of ammonia molecules, which are the starting materials of the reverse reaction, is constantly increasing, increasing the probability of the collisions that initiate the reverse reaction. We must also consider the effect of the heat energy released by the forward reaction. Since our gases are contained in a sealed vessel insulated against heat loss, the temperature of the mixture will rise as the forward reaction proceeds, since that reaction releases energy which disperses into the gas mixture. This temperature rise makes it increasingly likely that ammonia molecules will have sufficient kinetic energy to participate in the reverse reaction when they collide. In other words, it makes it increasingly possible for energy to disperse back into the internal structure of the chemicals and out of the heat of their surroundings.

This reaction we are considering is a rather complex one, and I am only highlighting the most important factors in this simple consideration, but the essential points are as follows:

at first, the natural tendency for energy to disperse towards an even distribution allows the forward reaction to proceed, releasing some of the energy stored in N_2 and H_2 as kinetic energy which disperses into the motions of all the molecules in the gas mixture,

but as the temperature of the gas rises and more ammonia molecules are produced it becomes increasingly likely that energy will disperse back *into* the lower energy NH_3 molecules and out of their high temperature surroundings.

In other words, the reaction eventually reaches a point at which it is *equally likely* that energy will disperse into the chemicals as out of them. This is the equilibrium point of the reaction, at which both forward and reverse reactions are still proceeding, but at equal rates. It corresponds to the point at which the energy within the system, both potential and kinetic, has become as evenly dispersed as it possibly can under the prevailing conditions.

So chemical reactions can have a great deal in common with what happens when a block of hot metal is dropped into cold water. At first the inevitable tendency of energy to disperse towards an even distribution causes a reaction in one direction to proceed much faster than the reverse process, just as heat at first moves out of the block of metal much faster than it moves in; but eventually an equilibrium point is reached at which energy has become as evenly dispersed as possible, and the forward and reverse processes are equally likely and so proceed at equal speeds. Having established the basic features of an 'equilibrium reaction', let's look at its behaviour a bit more, because there are other interesting things to observe.

Suppose we let our gas mixture settle down to its equilibrium point, and then we remove the thick layer of insulation around the vessel. What happens? We have opened up a route for heat energy to escape by dispersing into the cooler surroundings. Immediately, heat begins to escape, the gas mixture cools and so the forward reaction, which releases heat, becomes more likely than the reverse one, which absorbs heat. By allowing the gas mixture to cool we have upset the 'even distribution' of energy which had been set up, lowering the kinetic energy of the gas mixture and so encouraging the forward reaction which releases kinetic energy, and discouraging the reverse one which absorbs it. For a while the forward reaction proceeds much more quickly than the reverse one; but as it proceeds it is recreating the conditions which favour the reverse reaction, since it is pumping heat into the gas mixture and it is generating ammonia molecules. Eventually a new equilibrium point will be reached. This new equilibrium point will come when the gas mixture contains much more ammonia than it did at the original equilibrium point, and much less nitrogen and hydrogen, all of which allows the reverse reaction to proceed as fast as the forward one despite the fact that the temperature will be lower.

What if we then apply a flame to our vessel to heat up its gas? Again, the equilibrium position will adjust, but this time in the opposite direction. By pumping heat into the gas mixture we are encouraging the reverse reaction, which absorbs heat energy, much more than the forward reaction, which releases it. In other words, by pumping energy into the system we are encouraging the chemicals to react in the way that allows that energy to disperse into the chemicals of the system. As the reverse reaction proceeds, however, it is creating the conditions which encourage the forward reaction, because it is generating nitrogen and hydrogen

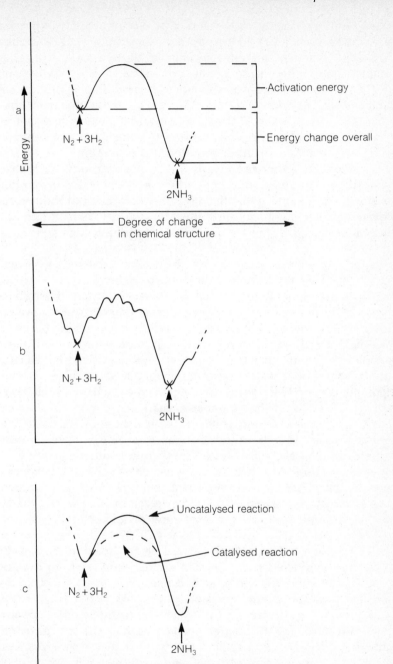

Figure 7.2 Energy profile diagrams of a chemical reaction (see text for details).

which are the starting materials of that reaction. Eventually a third equilibrium point is reached, at which forward and reverse reactions are again proceeding at equal rates and at which the hydrogen, nitrogen and ammonia are present in some new proportion.

The two big messages you should be learning from all this are, first, that chemical reactions are reversible; and secondly, that reactions do not inevitably proceed in the direction that generates lower energy products, but proceed in whatever direction the tendency of energy to disperse towards an even distribution will take them.

So reactions do move towards lower energy products, releasing energy into the surroundings, provided the energy of the surroundings is low enough to make that the most likely process; but they proceed towards higher energy products, absorbing energy from the surroundings, if the energy of the surroundings is sufficiently high to make that the most likely process.

Chemicals tend to react in a way which adjusts their energy levels to levels that are compatible with their surroundings.

We can make more sense of that last statement by introducing a new type of diagram, called an 'energy profile' diagram (see figure 7.2). Imagine we were able to work out exactly how much energy any chemical contained. We would know, in other words, the extent to which the arrangement of each chemical's nuclei and electrons involved the defiance of the electromagnetic force which we label 'potential energy', and the energy of movement which we call 'kinetic energy'. We would then be able to mark the energy level of each chemical against an energy scale. Such calculations can be made, but it is much easier to look at the 'difference' in energy levels between different chemicals, since we can measure this difference when it is released or absorbed as heat energy when one chemical is transformed into another. If we are only concerned about the differences in chemicals' energy levels, their relative energy levels in other words, then we can use a simple scale without any figures on it like the vertical axis of figure 7.2.

We can then write in a cross whose height up from the base line indicates the total energy embodied in the structure of, for example, one nitrogen molecule and three hydrogen molecules. We could then indicate how the energy level of these chemicals changes if their structure changes in any way by using the horizontal axis to represent, in a qualitative way, the degree of change which the structure of the chemicals is undergoing. The fact that we have placed the cross representing the combined energy of N_2 and H_2 at the minimum point of a steep energy 'valley' or 'well', indicates that to change their structure in any small way involves an increase in energy. So if the molecules became squashed or stretched or jolted out of their stable electronic structures in any way, their energy would rise.

The complete diagram illustrates nicely the potential for change embodied in the structure of N_2 and H_2, and also the resistance to that change. It shows that the nitrogen and hydrogen can, in principle, react to reach another stable 'energy well' of lower total energy by reacting to form ammonia. But it also shows us that in order for this major structural rearrangement to proceed, the chemicals will need to pass through intermediate states of much higher energy. So there is an 'energy barrier' preventing the reaction from proceeding, and it will only proceed if something disturbs the structure of the nitrogen and hydrogen molecules so much that they can surmount this energy barrier. In fact, as you already know, the barrier can be surmounted if the starting materials collide with such force that their structures are thrown into a temporary high energy chaos which can then settle down into the new lower energy stable structure of ammonia.

The energy difference between the bottom of the starting materials' energy well, and the top of the energy 'hill' facing them, is called the activation energy for the reaction. It is the energy needed to activate the reactants into reacting. The amount of energy released overall when the reaction proceeds is obviously the difference between the bottom of the starting materials well and the bottom of the products well.

The graph of figure 7.2(a) merely provides us with a very simple diagrammatic model that helps us to think about the overall energy changes accompanying chemical reactions. A more accurate model would show all of the rather shallow energy wells of the various high energy intermediate states which the reactants can adopt during their transition from nitrogen and hydrogen into ammonia. It would look more like figure 7.2(b). A three-dimensional model would be even better, with many different energy wells set at different depths into an undulating landscape like the side of a hill. Each energy well would represent some relatively stable, reasonably well balanced, electronic structure which the chemicals concerned could adopt; and the chemicals would be jolted from well to well by the energy they receive from their surroundings. Even the simplest two-dimensional model, however, makes several aspects of chemistry clearer to us.

Its message about chemicals moving between 'energy wells' is a general one. Any chemical which lasts long enough for us to discover it and study it must have a structure that is relatively resistant to change, even if the chemical stores a lot of energy within that structure. All chemicals are held together by chemical bonds, and it always takes an imput of energy to break these bonds, while energy is always given out when the bonds are formed. That, essentially, is the origin of the slopes at the sides of the energy wells. Movement up the slopes corresponds to breaking some or all of the bonds which hold the chemicals concerned together, while movement downwards to the bottom of the wells corresponds to the formation of these bonds. The chemical states represented by the tops of the slopes, by the peaks in an energy diagram, correspond to very unstable and therefore short-lived intermediate states in which all or some of the bonds of a chemical have been broken. If the chemical 'slides down the slope' back into the energy well it came from, then the old bonds are simply being remade; but there is also the possibility for it to slide down the other side of an 'energy hill', into a new well which corresponds to the formation of new bonds and new chemicals.

All chemical reactions involve the breaking and making of bonds. Existing bonds must be broken in order to allow the new bonds corresponding to the new arrangements of the chemicals to be formed. Breaking the bonds always requires an input of energy, while the formation of bonds always releases energy. The overall energy change of a reaction is determined by the net effect of all the separate bond-breaking and bond-making processes.

So it is important to realize that even when chemicals are reacting to form higher energy products, the final step in the reaction will involve a drop in energy as the chemicals fall down into an energy well, albeit one higher up the energy 'mountainside' than the one they came from. Conversely, even when chemicals are reacting to form lower energy products, the initial step in the reaction will involve a rise in energy, out of an energy well, as some existing bonds are broken in order to allow the new ones to form. These important points are simply and graphically conveyed by diagrams such as figure 7.2(a) and (b). Chemical reactions involve chemicals moving between energy wells in which they become trapped long enough for us to observe and study and exploit them. All of the chemicals we can see around us, in the natural world or stored in bottles on a laboratory shelf, are at the bottom of energy wells. The structures at the tops of the wells are the highly unstable and short-lived states which chemicals adopt as they move from well to well during reactions, but which never survive long enough for us to get a really good look at them. Chemists can manage a few fleeting glimpses of them, using sophisticated techniques capable of studying such short-lived structures, but the rest of us know them only by their effects on the sluggish and much more stable world we inhabit at the bottom of the wells.

The energy landscape around any chemical at the bottom of an energy well is not fixed. It can change as soon as any new chemical arrives or is mixed in to create new possibilities for reaction. Almost all chemical reactions, for example, can be made to proceed more quickly by the addition of some appropriate 'catalyst'.

A catalyst is a substance which speeds up a chemical reaction while itself remaining unchanged overall; and catalysts achieve their effects by opening up new routes for a reaction to follow, routes which involve a lower 'energy hump' (i.e. a lower activation energy) than the routes otherwise available to them. Catalysts can work in many different ways, but one of the most common mechanisms can be illustrated using the reaction in which nitrogen and hydrogen form ammonia.

In the absence of any outside help this reaction proceeds far too slowly to be industrially useful, but if a catalyst composed of finely divided iron and some other chemicals is added to the reaction chamber, the reaction can speed up dramatically. Essentially, the iron is believed to offer a chemical surface onto which the nitrogen and hydrogen molecules can become adsorbed in a way which encourages them to react together to form ammonia. It provides a meeting place for the reactants, whose chemical nature happens to encourage their successful reaction. Using such a catalyst, the reaction between nitrogen and hydrogen proceeds fast enough to serve as the basic feed reaction of the world fertilizer industry, generating valuable ammonia in plentiful abundance. When the energy profile of the reaction in the presence of the catalyst is superimposed on figure 7.2(a) to yield figure 7.2(c), we see that the presence of the iron has opened up an alternative pathway by which the reaction can proceed, a pathway involving a much lower activation energy, and therefore one which many more molecules of nitrogen and hydrogen are able to follow. In the presence of the catalyst, a significantly larger proportion of the nitrogen and hydrogen molecules possess the energy needed to react to produce ammonia, simply because the presence of the catalyst has made that energy level considerably lower.

It is important to realize that the catalyst does not only speed up the forward reaction, it also speeds up the reverse reaction, since it cannot lower the forward activation energy without similarly lowering the reverse activation energy. All that the catalyst does is reduce the time it takes for a reaction to reach its equilibrium position, at which forward and reverse reactions proceed at the same rates—it does not affect the nature of that equilibrium position at all. The problem with the industrial utilization of the ammonia-forming reaction was not the final equilibrium position of the reaction, it was the time it would take the reaction to reach that position. The catalyst vastly reduces the time it takes for a sample of nitrogen and hydrogen to generate significant quantities of ammonia, and that is why it is so useful.

You have now met most of the fundamental principles of chemistry, principles whose power you will be able to appreciate as you read through the remaining chapters of this book. Each of these chapters examines one interesting chemical system or topic—mineral, living, industrial or medical. Before we proceed with these explorations of some of the fascinating facets of the world of real chemical systems, however, we should briefly take stock. The chemicals around and within us are composed of atoms of matter, which are themselves composed of different arrangements of the sub-atomic particles: protons (mass = 1 amu, electric charge = +1), neutrons (mass = 1 amu, no overall charge) and electrons (mass = 0.0005 amu, electric charge = −1). The protons and neutrons are bundled together into a tiny central core, the nucleus, of an atom, while the electrons occupy much larger orbitals around the nucleus. Each orbital corresponds to a particular energy level for any electron within it, the energy being sufficient to keep the electron from being pulled further towards the nucleus by the electric force.

The electric force which causes like charges to repel one another and unlike charges to be attracted towards one another is the fundamental force behind all of the pulling and pushing and changing that goes on in the chemical microworld; but this force is resisted by the phenomenon of energy, and the direction of chemical change is guided by the automatic and inevitable tendency of this energy to disperse towards a more even distribution.

As a result of the flow of energy through the microworld, all the particles of chemistry are moving about and colliding with one another. One result of these collisions is to make chemical reactions proceed, reactions in which atoms can become bonded together into compounds in which the atoms are held together by covalent and polar covalent bonds, or in which electrically charged ions are created from atoms or molecules and then drawn towards one another by the electric force.

All chemical change involves particles (atoms and/or molecules and/ or ions) colliding in ways which allow the competing strains of the electric force and energy to rearrange the particles into new types of substance, all guided by the inevitable dispersal of energy towards a more even distribution overall.

So chemistry is a frantic dance of particles in which some participants are drawn towards one another while others are forcefully repelled, with the whirling energy of the dance defying these ordering forces as it spreads across the dancefloor.

Box 7.1 Chemical shorthand

Chemists represent the chemicals of the earth and the reactions they participate in using chemical formulae which are combined into chemical equations.

A chemical formula is simply a symbolic representation of which atoms or ions are present in a compound, and the proportions in which they are present. The basic 'letters' of the chemical alphabet are the symbols for the elements shown in the periodic table. Some combination of these symbols along with numbers to represent the relevant proportions, are the only constituents of all chemical formulae.

H_2O, for example, is the formula for water. It illustrates the rule that simple numerical subscripts in formulae refer to the element whose symbol immediately precedes the subscript. So in this case the '2' belongs to the 'H', and indicates that two hydrogen atoms are found for every one oxygen atom in water. Water, in fact, consists of discrete molecules each containing two hydrogen atoms and one oxygen atom.

Another example of this system can be seen in the formula of ethanol, the intoxifying 'alcohol' most of us like to consume from time to time, which is C_2H_6O, indicating that alcohol molecules each contain two carbon atoms, six hydrogen atoms and one oxygen atom.

Some compounds have two types of formula: a molecular formula and also an 'empirical' formula. The chemical 'octane' for example, a constituent of petrol, consists of molecules containing eight carbon atoms and eighteen hydrogen atoms, so its molecular formula is C_8H_{18}. The empirical formula of a compound is merely the *simplest* way of representing the proportions of atoms present in the compound, using whole numbers and not worrying about the molecular structure of the compound. The simplest whole number formula representing the chemical content of octane is clearly C_4H_9, so this is its empirical formula. The molecular and empirical formulae of many chemicals are identical to one another, but for many others they are not.

The chemicals considered so far have been covalently bonded compounds which therefore exist in the form of discrete molecules. A complication emerges when we consider the formulae of ionic compounds.

Simple ionic compounds, composed of ions derived directly from atoms which have lost or gained electrons, are easily dealt with as above. Thus sodium chloride, consisting of positive sodium (Na^+) ions and negative chloride (Cl^-) ions in a 1:1 ratio, is represented by the formula Na^+Cl^-, or simply NaCl (the charges on the ions are often not indicated). Equally, calcium fluoride, consisting of calcium (Ca^{2+}) and fluoride (F^-) ions in a 1:2 ratio can be represented as CaF_2.

Many ions, however, are formed from a combination of more than one type of element. Examples of such 'compound' ions are hydroxide ions (OH^-), consisting of a hydrogen atom bonded to an oxygen atom and with an overall single negative charge; ammonium ions (NH_4^+) carbonate ions (CO_3^{2-}), consisting of a carbon atom bonded to three oxygen atoms and with an overall double negative charge. When multiples of these ions are used in the formula of a compound, the formula of the ions must be enclosed within brackets to make it clear what the various numbers refer to. Numerical subscripts after such brackets refer to the entire contents of the brackets.

The formula of ammonium carbonate, for example, is $(NH_4)_2CO_3$, indicating that two ammonium ions are present for every one carbonate ion. The formula for calcium hydroxide is $Ca(OH)_2$, indicating two hydroxide ions for every calcium ion, and so on . . .

A chemical equation merely lists the chemicals which take part in a chemical reaction, and the products that are formed from them, using chemical formulae to represent the individual chemicals and large numbers before these formulae to indicate the proportions in which the individual chemicals react. Any number before the formula of a compound in an equation applies to the entire compound, so that 2NaOH really means $2 \times [NaOH]$ and not just two sodium ions and one hydroxide ion. Similarly $3K_2CO_3$ refers to three 'formula units' of potassium carbonate, which will contain a total of six potassium ions, and three carbonate ions (which will contain three carbons and nine oxygens overall). Check it out to make sure you understand!

We have already seen some simple chemical equations such as the one used to describe the creation of ammonia from hydrogen and nitrogen:

$$N_2 + 3H_2 \rightarrow 2NH_3$$

Even such a simple equation embodies all the essential principles behind all equations. We create the equation by first listing the correct formulae of the compounds involved, separated into reactants and products. We must add any numbers neccesary to ensure

Group		
1	2	
1 H Hydrogen		
3 Li 2.8.1 Lithium	4 Be 2.2 Beryllium	
11 Na 2.8.1 Sodium	12 Mg 2.8.2 Magnesium	
19 K 2.8.8.1 Potassium	20 Ca 2.8.8.2 Calcium	
37 Rb 2.6.18.8.1 Rubidium	38 Sr 2.8.18.8.2 Strontium	
55 Cs 2.8.18.18.8.1 Caesium	56 Ba 2.8.18.18.8.2 Barium	
87 Fr 2.8.18.32.18.8.1 Francium	88 Ra 2.8.18.32.18.8.2 Radium	

Transition metals:

21 Sc 2.8.9.2 Scandium	22 Ti 2.8.10.2 Titanium	23 V 2.8.11.2 Vanadium	24 Cr 2.8.13.1 Chromium	25 Mn 2.8.13.2 Manganese	26 Fe 2.8.14.2 Iron	27 Co 2.8.15.2 Cobalt	28 Ni 2.8.16.2 Nickel	29 Cu 2.8.18.1 Copper	30 Zn 2.8.18.2 Zinc
39 Y 2.8.18.9.2 Yttrium	40 Zr 2.8.18.10.2 Zirconium	41 Nb 2.8.18.11.2 Niobium	42 Mo 2.8.18.13.1 Molybdenum	43 Tc 2.8.18.14.1 Technetium	44 Ru 2.8.18.15.1 Ruthenium	45 Rh 2.8.18.16.1 Rhodium	46 Pd 2.8.18.18.0 Palladium	47 Ag 2.8.18.18.1 Silver	48 Cd 2.8.18.18.2 Cadmium
57 La 2.8.18.18.9.2 Lanthanum	72 Hf 2.8.18.32.10.2 Hafnium	73 Ta 2.8.18.32.11.2 Tantalum	74 W 2.8.18.32.12.2 Tungsten	75 Re 2.8.18.32.13.2 Rhenium	76 Os 2.8.18.32.14.2 Osmium	77 Ir 2.8.18.32.15.2 Iridium	78 Pt 2.8.18.32.17.1 Platinum	79 Au 2.8.18.32.18.1 Gold	80 Hg 2.8.18.32.18.2 Mercury
89 Ac 2.8.18.32.18.9.2 Actinium									

Main groups (3–8):

3	4	5	6	7	8
					2 He 2 Helium
5 B 2.3 Boron	6 C 2.4 Carbon	7 N 2.5 Nitrogen	8 O 2.6 Oxygen	9 F 2.7 Fluorine	10 Ne 2.8 Neon
13 Al 2.8.3 Aluminium	14 Si 2.8.4 Silicon	15 P 2.8.5 Phosphorus	16 S 2.8.6 Sulphur	17 Cl 2.8.7 Chlorine	18 Ar 2.8.8 Argon
31 Ga 2.8.18.3 Gallium	32 Ge 2.8.18.4 Germanium	33 As 2.8.18.5 Arsenic	34 Se 2.8.18.6 Selenium	35 Br 2.8.18.7 Bromine	36 Kr 2.8.18.8 Krypton
49 In 2.8.18.18.3 Indium	50 Sn 2.8.18.18.4 Tin	51 Sb 2.8.18.18.5 Antimony	52 Te 2.8.18.18.6 Tellurium	53 I 2.8.18.18.7 Iodine	54 Xe 2.8.18.18.8 Xenon
81 Tl 2.8.18.32.18.3 Thallium	82 Pb 2.8.18.32.18.4 Lead	83 Bi 2.8.18.32.18.5 Bismuth	84 Po 2.8.18.32.18.6 Polonium	85 At 2.8.18.32.18.7 Astatine	86 Rn 2.8.18.32.18.8 Radon

Lanthanides:

58 Ce 2.8.18.20.8.2 Cerium	59 Pr 2.8.18.21.8.2 Praseodymium	60 Nd 2.8.18.22.8.2 Neodymium	61 Pm 2.8.18.23.8.2 Promethium	62 Sm 2.8.18.24.8.2 Samarium	63 Eu 2.8.18.25.8.2 Europium	64 Gd 2.8.18.25.9.2 Gadolinium	65 Tb 2.8.18.27.8.2 Terbium	66 Dy 2.8.18.28.8.2 Dysprosium	67 Ho 2.8.18.29.8.2 Holmium	68 Er 2.8.18.30.8.2 Erbium	69 Tm 2.8.18.31.8.2 Thulium	70 Yb 2.8.18.32.8.2 Ytterbium	71 Lu 2.8.18.32.9.2 Lutetium

Actinides:

90 Th 2.8.18.32.18.10.2 Thorium	91 Pa 2.8.18.32.20.9.2 Protactinium	92 U 2.8.18.32.21.9.2 Uranium

Electron arrangement (in this case, two electrons in first shell, eight in second shell, three in third shell)

Figure 7.3 A periodic table including the electron arrangement of each type of atom, and the numbers of the 'main groups'. The electron arrangements list the number of electrons in each shell, starting with the inner shell and moving outwards. Only naturally occurring elements are shown.

that the equation is chemically balanced, in other words that no atoms are created or destroyed during the course of reaction. In balancing an equation we must restrict ourselves to varying the proportions in which the participating chemicals combine, resisting the temptation to alter the formulae of any of the compounds in order to 'make things fit'. Nature balances her chemical reactions automatically, since balanced reactions are the only ones possible. We sometimes have to scratch our heads and fiddle with the numbers for quite a while before we work out how she manages it!

Chemical equations are simply a shorthand way of representing information about a reaction which could be put down in words. There is nothing mystical or magical about them—they simply allow chemists to communicate in one line information which might otherwise take several paragraphs.

Box 7.2 The search for stability

Out of the 92 naturally occurring elements of the periodic table, there are six that do not really do anything, that do not, in other words, tend to participate in chemical reactions. This chemical inactivity is actually very revealing.

The six in question are the so-called 'noble gases'—helium, neon, argon, krypton, xenon and radon—which you can find in the rightmost group (vertical column) of the periodic table of figure 7.3. The name comes from their chemical aloofness, akin to that of the nobility. The noble gases seem persistently reluctant to mix and react with their fellow elements, a reluctance recognized more obviously in their alternative name of the 'inert' gases.*

You have been told that atoms react together in ways which cause rearrangements of their electrons and often tend to trap the electrons in new, more stable, electron arrangements at the bottom of deeper energy wells. So the reluctance of the noble gases to react seems to imply that their electron arrangements are already very stable ones, already occupying deep energy wells. This is indeed the case, for there is something rather special about the electron arrangement of all of them.

---

*These so-called 'inert' gases are not in fact completely inert. They can be induced to participate in some unusual and exotic reactions if a considerable effort is made to make them do so.

Helium has two electrons, which both fit into the single spherical oribital of the first electron shell (look back to pages 41–8 if you need to be reminded about electron shells and subshells). This electron arrangement means that helium atoms possess a completely filled outer electron shell (since they only have one shell, containing the one orbital, and there is only room for two electrons in that orbital).

The next noble gas, neon, has a total of 10 electrons, so two fit into the inner electron shell while eight fit into the second shell. These eight outer electrons are distributed between the two subshells of this second shell, which are again completely filled, since the first subshell contains one orbital and the second contains three, making room for eight electrons overall. So neon too has a completely filled outer electron shell.

The situation with the remaining four noble gases is similar, but not quite so logically neat. Notice from figure 7.3 that all four have eight electrons in their outer electron shell, just like neon. So the arrangement of their outer electron shells is the same as neon's, each consisting of a filled first subshell and a filled second subshell. The complication is that argon, krypton, xenon and radon also have completely empty further subshells available, but we appear to have identified a common factor which may underlie the unusual stability and unreactivity of the noble gases. The electron arrangement of the outer electron shell of each of them consists of either a completely filled electron shell (helium and neon), or completely filled first and second subshells with any other subshells being empty (argon, krypton, xenon, radon). The possession of completely filled shells or subshells of outer electrons corresponds to neatly symmetrical arrangements of electrons which are noticably more resistant to change, or in other words more stable, than other arrangements. That is essentially why the noble gases are so reluctant to react—because their electrons are already in the sort of stable arrangements which chemical reactions tend to take other atoms towards. The rule revealed to us by the noble gases, that having filled shells and subshells of outer electrons is an unusually stable situation, is actually only the most prominent aspect of a broader rule which says that either filled or half-filled shells or subshells correspond to unusually stable electronic structures. It seems to be the overall symmetry of these configurations which gives them their unusual stability.

These observations about the unusually stable electron structures of the noble gases are very revealing because they allow us to identify a very common pattern in the chemical reactions between other atoms which do not possess such stable electron arrangements. When the reactions of many other atoms are studied, we find that these reactions tend to leave the atoms with modified electron arrangements which are much more like the stable configurations of the noble gases. It seems, in other words, as if other atoms often participate in reactions which allow them to attain the very stable 'noble gas' electron arrangements, with the reacting chemicals often being trapped in these stable arrangements because the energy needed to jolt them back out of them so often disperses away during the course of a reaction.

The most common noble gas outer electron structure, involving eight outer electrons occupying filled first and second subshells of the outer shell, is known as the 'stable octet' structure. This is the outer electron structure of all the noble gases except helium (which has insufficient electrons to adopt it), and it is one of the most stable electron arrangements available for atoms to become trapped in when they participate in chemical reactions.

The concept of atoms reacting 'in order to attain' a stable octet of outer electrons is so pleasingly simple that some elementary chemistry books raise it to the status of a great law of chemical reactivity. To do this is very misleading. It can create the impression that the only reason chemicals react is in order to achieve this one particularly stable electron arrangement (or the helium arrangement if there are insufficient electrons to form an octet). If you take the trouble to examine the electronic structure of a wide variety of compounds you can soon discover many exceptions to the octet rule. This is because it is not a rule at all, but merely a guide to one particularly common pattern of chemical reactivity. Many chemical compounds contain atoms with relatively stable outer electron arrangements which do not resemble the stable noble gas arrangements. So while the electron arrangements of the noble gases are particularly stable arrangements which many atoms adopt as they react, there are many other stable arrangements as well. The stable octet represents just the deepest of several low energy wells available to electrons as they become rearranged during chemical reactions.

Having emphasized the weaknesses of this widely known simplifying concept of chemistry, let me make its strengths clear to you with some examples in which it works very well. Sodium atoms have a single outer electron, as you can see from figure 7.3; but if they could lose this electron during a chemical reaction a sodium ion would be created whose electron arrangement would now be identical to that of a neon atom, including a 'stable octet' of electrons in what would become the outer shell. This is essentially why sodium atoms tend to react to form Na^+ ions, because by doing so they attain a particularly stable electronic structure. Similar logic applies to all the other elements of the periodic table's first group, since they too possess one outer electron above an inner core of electrons identical to the electron arrangement of a noble gas.

At the other extreme of the periodic table, we find elements such as chlorine, all with seven outer shell electrons and therefore needing to gain just one more in order to complete a stable octet. This explains why the elements of group 7 of the table tend to react by gaining electrons to form singly charged negative ions such as Cl^-.

So the reaction between sodium and chlorine, for example, to form the ionic compound sodium chloride (Na^+Cl^-) can be rationalized as due to an electron transfer, from a sodium atom to a chlorine one, which creates two ions which both then possess a stable electron arrangement similar to a noble gas. I should point out, however, that the electron arrangement of these ions in a free unbonded state is nevertheless a higher energy one than the corresponding free atomic state. The combined energy of a free Na^+ ion and a free Cl^- ion, in other words, is higher than that of a free Na atom and a free Cl atom; but if the oppositely charged ions then come together to be joined by an ionic bond, then there is a large decrease in energy which more than compensates for the energy input needed for the original electron transfer to take place. (The ions can also be stabilized independently by a fall in energy when they interact with water, as you will learn in chapter 10).

As you have worked through this box you should have noticed one of the fundamental structural features of the periodic table. Elements in the same group of the table usually have the same number of outer electrons; and since the outer electron arrangements of atoms are the major determinants of their chemical activities, atoms in the same group of the periodic table tend to behave in chemically very similar ways.

We can see this concept confirmed if we move on to group 2. All of the group 2 atoms have two outer electrons above an inner electron core identical to one of the noble gases. So the group 2 elements tend to lose two electrons to form doubly charged positive ions when they react to form ionic compounds. The unusual stability of the 'noble gas' electron core of magnesium atoms, for example, explains why they react to form Mg^{2+} ions, and not Mg^+ or Mg^{3+}. So the formula of the compound formed when magnesium reacts with chlorine is $Mg^{2+}Cl_2^-$ (or simply $MgCl_2$), with each magnesium atom providing an electron for each of two chlorine atoms, leaving all three ions concerned with the stable outer electron arrangement of a noble gas.

Similarly, elements of group 6 have six outer electrons, so the easiest way for them to attain a noble gas structure is to acquire two electrons to form doubly charged negative ions. This is why oxygen atoms, when reacting to form ionic compounds, form O^{2-} ions, and sulphur atoms form S^{2-} ions and so on. So the formula of sodium sulphide is $Na_2^+S^-$ (or simply Na_2S); and the formula of calcium oxide is $Ca^{2+}O^{2-}$ (CaO); and the formula of potassium oxide is $K_2^+O^{2-}$ (K_2O).

So far we have looked only at ionic compounds, and the way in which the formation of many ions can be rationalized by the alteration of atoms to leave them with a stable, noble gas electron structure. The idea can be applied to compounds held together by covalent bonds as well, although it only works clearly using the valence bond approach to bonding discussed in chapter 6. An oxygen atom, for example, has six outer electrons, two of which are found in half-filled orbitals and are therefore available to overlap with the half-filled orbitals of other atoms to form bonds according to the reasoning of the valence bond approach to bonding. In a molecule of water, these half-filled orbitals of oxygen are envisaged as overlapping with the single half-filled orbitals of the hydrogen atoms, to generate bonding orbitals as shown in figure 6.9. If you then count the electrons arranged around the oxygen atom, you find that it is now surrounded by eight of them. Although four of these electrons are shared with the hydrogen atoms, this electron arrangement around the oxygen atom in the molecule is certainly closer to a stable octet than the arrangement around a lone oxygen atom. Similarly, each hydrogen atom now has a share of two electrons, making the electron arrangement around the hydrogen atoms closer to that of an unreactive helium atom than the arrangement around lone hydrogen atoms. This sort of reasoning can be used to rationalize the reactions of a great many atoms when they combine to form molecules held together by covalent bonds. In many cases the electron arrangements around the atoms seem to be much more like one of the noble gas arrangements, compared to the situation in the free atoms.

These ideas certainly do not work for all covalently bonded compounds, nor indeed for all ionic compounds, and they become less obviously useful when the probably more accurate molecular bonding approach to chemical bonds is adopted.

The single fundamental point about the electron arrangements which form all chemical bonds is that they represent stable structures for the chemicals concerned, structures which are at the bottom of energy wells, and which chemicals therefore tend to 'fall down into' and 'jump between' as they participate in chemical reactions. The idea that the electron arrangements corresponding to the bottom of these energy wells often closely match the electronic structures of the noble gas atoms is a useful guide in many cases, but it is an approximate and fallible guide. It is of only limited value compared to the more accurate idea of a wide range of energy wells, each corresponding to the unique situation of one particular chemical and the bonds that hold it together.

8 Fire

A metal pipe thrust into the sea floor has punctured a reservoir of 'natural gas'—largely 'methane'—composed of hydrogen and carbon atoms which millions of years ago were part of a living thing. The gas rushes upwards out of this hole in its rocky tomb and flows through many miles of pipework, across the seabed and then beneath the mountains and valleys of Scotland until it reaches my home. It hisses from the top of my cooker until I press the ignition button, causing an electric spark (a rush of electrons from molecule to molecule within the gas/air mixture) to ignite it into flame. I have started a fire which will release some energy which was first captured from the sun so long ago. I will use it to boil the water for my coffee.

Most fires are chemical reactions in which some substance is combined with oxygen. The various atoms of the fuel end up linked to oxygen atoms in the form of chemicals such as carbon dioxide (CO_2), water (H_2O), sulphur dioxide (SO_2), nitrogen dioxide (NO_2) and so on. The energy embodied within the structure of these 'oxidized' products of combustion is always less than the energy of the reactants—the fuel and the oxygen which combined with it. As the reaction proceeds, the excess energy of the reactants disperses out into the environment as heat and light, and a little of it in the form of the waves of air compression which cause the sound of the burning flame. Humans have learned how to use the energy released from flames, to keep us warm, to scare away the predatory animals of the night, to drive the machines which power our modern civilization, to blast us away from the earth on journeys through space, and to blast many of our fellow humans into quick oblivion—chemically, many explosions are merely very fast fires.

We should examine the chemistry of fire in a little more detail. The reactants of the fire which boils the water for my coffee are methane (CH_4) and oxygen (O_2). These combine in the overall proportions of 1:2 to yield carbon dioxide and water, as described by the following equation:

$$CH_4 + 2O_2 \rightarrow CO_2 + 2H_2O$$

So methane and oxygen flow into the flame, while water vapour and carbon dioxide rise out from its top; but what actually happens in the bright blue high energy chaos of the flame itself?

There is certainly not a series of simple collisions between each methane molecule and two oxygen molecules to generate one molecule of carbon dioxide and two of water. The heat of the flame (initially just the heat of the igniting spark) forces the reacting molecules to collide with such violence that they are split into free atoms or ions or small and unstable molecular fragments. These atoms and/or ions and/or fragments of molecules are the first high energy intermediates which then participate in a series of further reactions, each yielding new intermediate products until, finally, molecules of CO_2 and H_2O are generated which fall into their low energy wells and react no further.

The precise internal mechanism of the reaction, as with most reactions, is a horribly complicated high energy mess which is difficult to study in detail. Many such reaction pathways have been studied in some detail, however, and they reveal a very complex picture. The simplest fire of all, between hydrogen and oxygen, is known to involve at least 14 individual chemical steps. The burning of a methane flame probably involves many more, while the burning of any complex substance such as wood, which contains many thousands of different chemical compounds, must involve many thousands if not millions of distinct chemical steps. Overall, however, everything is blissfully simple. For every one methane molecule in my flame, two oxygen molecules from the air are consumed to generate one molecule of carbon dioxide, two molecules of water, plus a lot of energy; and it all happens because chemicals collide in ways which allow the electromagnetic force to pull their nuclei and electrons into new arrangements, with lower energy arrangements being preferentially trapped by the dispersal into the environment of the energy which would be needed to push the chemicals back into higher energy states.

Much of the energy comes out as heat. It appears, in other words, in the form of the violent motion of the particles in the flame—motion which jostles into the bottom of my kettle and through the assembly of moving metal atoms to appear in the increasingly violent motion of the molecules of warming water. The dispersal of heat energy via collisions between particles in motion takes it from the heart of the flame into the water which will eventually warm the contents of my stomach.

Some of the energy also comes out as light. Such a large quantity of energy is released during the reaction that it pushes many of the particles in the heart of the reaction into 'excited states' in which some of their electrons are in higher energy orbitals than they need be. These electrons, in other words, are occupying high energy orbitals despite the fact that empty lower energy orbitals are available. Such high energy electrons eventually fall down into the lower energy orbitals, releasing energy in the form of electromagnetic radiation as they go. Some of this radiation is in the visible part of the electromagnetic spectrum, so we see it as the light of the flame, although invisible radiations are released as well. Infrared radiation, remember, is a form of electromagnetic radiation which can be regarded as radiant heat. It is released when the bonds within chemicals lose some of their energy of vibration, and is absorbed by other chemicals to increase the vibration of their bonds. The internal vibrations of chemicals are part of the phenomenon we call heat, which includes all forms of kinetic energy (energy of motion) held by chemicals. The release of infrared radiation is a second major way in which heat energy can disperse, in addition to the physical conduction of heat considered earlier. So invisible 'infra-red' radiations allow the heat of a flame to warm us directly, without requiring any conduction or convection involving intermediate particles, by travelling outwards at the speed of light until they are absorbed by our particles to increase their energy of motion.

A fire is a runaway chemical reaction which consumes fuel as long as there is fuel to burn, and which throws out energy in the form of heat and light and other electromagnetic radiations. As the heat energy disperses from the heart of the fire, some of it provides the necessary activation energy needed to jolt some more of the unburnt fuel into burning.

Every day of your life you make great use of the energy released by fire, or by some very fast fires we call explosions. The energy of fire boils your water and cooks your meals, for even if you cook by electricity most of that electricity is generated from the energy released from burning coal or oil. So every piece of mains-powered electrical equipment you use during the day is also largely driven by the power station fires which boil the water which makes the steam which turns the turbine blades of the generators which convert the energy of movement into the energy of electric current. The fast burning of petrol provides the explosive force to drive your car on your daily journeys to and from work, quickly releasing the heat energy which causes the rapid expansion of burning gas which pushes at the pistons in the engine's cylinders and ultimately turns the car wheels. Every second of the day and night your activities and your comfort are probably in some way dependent on the energy released from fires. In primitive societies fires became the centrepiece of each village and each home, and since such

times their central importance to all our lives has remained undimin-
ished, though it is often now much less obvious. Although nowadays
we may rarely see any flames, the effects of the energy released by fire
are all around us.

To be honest, however, I have presented you with a rather simplistic
view of the chemistry of fire. I have assumed that every methane
molecule, for example, is able to react with the two oxygen molecules
needed to convert it into carbon dioxide and water. I have assumed, in
other words, that a plentiful excess of oxygen has been available to
ensure that the burning of the methane proceeds 'to completion'. For
many fires, burning methane or any other fuel, this will not be the case,
and a deficient supply of oxygen can cause the fuel to be incompletely
burned. If methane is ignited in a poor supply of oxygen, it will not burn
with the roaring blue flame of a cooker, but with a quieter and yellow
'sooty' flame. We call it a sooty flame because soot, which is simply
carbon, is one of the products of the incomplete combustion of methane
and of all compounds containing carbon. The carbon atoms which form
soot can be released from methane molecules when the hydrogen
atoms of the molecules combine with the available oxygen to form
water vapour, leaving no oxygen available to react with the carbon
which therefore rises up from the flame as soot. Another possibility is
for the carbon to combine with oxygen to form carbon monoxide (CO)
rather than carbon dioxide. Such partial oxidation of carbon atoms in
petrol is the source of the poisonous carbon monoxide that issues from
motor exhausts. If enough oxygen were available to react with the
petrol completely, then the carbon atoms of the petrol would all be
converted into the completely oxidized form of carbon dioxide.

So in the chemistry of fire we have uncovered a major universal
complication of chemistry: only if the reactants happen to be present in
the precise proportions indicated by the reaction equation, and only if
they all then fully react, will the equation represent what really
happens. If some or all of the reactants are more abundant than the
equation requires them to be, then that excess will remain unreacted
and unchanged, or else some new more complex mixture of reactions
might occur which yields different products from those indicated by
the simple ideal overall equation. If some methane is ignited with too
little oxygen for it to be converted fully into carbon dioxide and water,
for example, some of the methane molecules will be fully oxidized into
carbon dioxide and water, some will be partially oxidized into carbon
monoxide and water, some will be partially oxidized into carbon (i.e.
soot) and water, while some may remain as unreacted methane.

Any simple equations written by chemists to summarize chemical reactions normally summarize only the aspects of the reactions they are most interested in, or which are most dominant. Such equations usually neglect to take account of the inevitable complexities due to the portions of the reactants which do not react fully, or which participate in minor 'side' reactions to yield alternative products, or which remain completely unreacted because there is an excess of them over their fellow reactants. The writing of chemical equations is a simplifying process which allows us to concentrate on the dominant or most interesting effects of chemical processes which are almost always more complex and messy than our equations make them seem.

9 Air

The air around you is a mixture of molecules and atoms, held down by the gravitational force and held up, in other words raised into a 200 mile high atmosphere rather than squashed against the surface of the earth, by the rapid motion of its particles. This motion causes them to bounce off one another and the earth to form a mass of jiggling particles which is thickest at ground level and steadily attenuates all the way up to the near vacuum of space.

Most of the air (78 per cent) consists of nitrogen in the form of N_2 molecules. The two atoms of each molecule are held together by three covalent bonds to form an extremely stable structure which requires a great deal of energy to break it apart. The energy of a thunderstorm can break it up, however, causing it to react with oxygen in the air to form compounds such as nitric oxide (NO); and when it does react a lot of energy can be released. Nitrogen occupies a deep well in the energy mountainside, making it very stable, but it is a well that is quite high up the mountainside.

The second main constituent of the atmosphere, oxygen (O_2), also contains a lot of energy embodied within its structure. We have already considered how energy is released when oxygen combines with many chemicals, such as methane and petrol, and we have looked at some of the uses to which that energy can be put. The oxygen we breathe into our bodies is distributed, via the bloodstream, to all the cells of the body, and within these cells it reacts with chemicals derived from our food to release the energy which keeps us warm and which powers all of the energy-requiring processes of life. Molecules of sugar (i.e. sucrose—$C_{12}H_{22}O_{11}$), for example, dissolved in our coffee or tea, can eventually be combined with oxygen to generate carbon dioxide (CO_2) and water (H_2O) as wastes:

$$C_{12}H_{22}O_{11} + 12O_2 \rightarrow 12CO_2 + 11H_2O$$

Our cells utilize the energy released by this overall reaction to power other energy requiring reactions, while we breathe the waste carbon dioxide into the atmosphere and can get rid of excess water in urine, or sweat, or as water vapour in our breath.

The reaction by which our bodies extract energy from sugar is, overall, the same reaction as the burning of sugar, but there are no sugary fires and flames within our cells. Instead, the same overall reaction as burning is achieved in many small controlled steps, allowing the energy to be released gradually, and then stored and utilized to sustain us rather than burn us to death.

Oxygen molecules account for just under 21 per cent of dry air (the atmosphere contains a variable amount of water vapour), so once the oxygen and the nitrogen (78 per cent) are taken into account there is little room left for anything else—a mere one per cent.

Most of that one per cent consists of atoms of argon, which accounts for about 0.9 per cent of the atoms of dry air. These argon atoms do not do much because they are very stable atoms indeed. Argon is one of the noble gases whose atoms already possess such a stable electron arrangement they have very little tendency to participate in reactions which would disturb that arrangement.

We have considered only three chemicals, and yet we have accounted for 99.9 per cent of the chemicals in dry air. Yet there are many thousands if not millions of different chemicals present in air. Obviously there must only be very small amounts of these other chemicals in the air, compared to its nitrogen, oxygen and argon, but these other 'trace' gases are nevertheless very significant indeed. They both preserve and threaten all of our lives, so we should certainly take a look at some of them.

Carbon dioxide (CO_2) is being pumped into the atmosphere at an incredible rate. Every fire, whether a bush fire, a forest fire, a coal or oil fire within a power station or one of the millions of tiny fires in homes, motor car cylinders and so on, pumps out the carbon dioxide generated by the oxidation of the carbon atoms within fuel. Every animal on earth, with every breath, releases some carbon dioxide generated by the oxidation of its food. So, natural fires, the many fires of industry and commerce and the slow flameless 'fires' of life make the surface of the earth a giant carbon dioxide producing machine.

Yet carbon dioxide is a relatively minor constituent of the atmosphere, at about 0. 03 per cent of dry air. This is largely because the plant life of the earth is busy absorbing carbon dioxide from the atmosphere almost as quickly as it is dumped into it. The water of the seas and oceans acts as another carbon dioxide 'sink', holding huge quantities of dissolved carbon dioxide which is gradually incorporated into sedimentary rocks composed of such compounds as calcium carbonate ($CaCO_3$) and magnesium carbonate ($MgCO_3$); but plants play the major and most immediate role.

Plants take up carbon dioxide and use it as a raw material which they need to grow and which, indirectly, is needed to sustain almost all life on earth. Essentially, plants reverse the chemical combustion of sugars, by combining carbon dioxide with water to generate substances such as glucose ($C_6O_{12}O_6$) and sucrose ($C_{12}H_{22}O_{11}$), along with oxygen gas. The sugars are then used as the basic carbon-containing raw materials of plant life. Notice that this must be an *energy-requiring* reaction, since the reverse reaction of combustion releases lots of energy. The reaction will only proceed in the energy-requiring direction if the energy needed to make it proceed disperses automatically into the reactants. The required energy comes in the form of sunlight, and it disperses into plants and is trapped within them during the complex biochemical process known as 'photosynthesis' (see chapter 14). Although photo-synthesis is an incredibly complex process involving hundreds of individual chemical steps and many different chemicals, its overall effects amount to the simple conversion of carbon dioxide and water into sugars and free oxygen gas.

So fires and animals (and also plants, during their non-photosynthetic activities) release carbon dioxide into the air, while plants absorb it from the air and incorporate it into new plant growth; and note also that while fires and animals consume oxygen gathered from the air, plants release it into the air. We are looking at important parts of two basic chemical *cycles* of the earth—the carbon cycle and the oxygen cycle. The earth can be regarded as a complex self-sustaining chemical machine, a machine which uses up particular chemicals in some processes but releases them again in others, with the whole machine powered by the energy arriving on earth from the sun and the heat energy constantly seeping up from the planet's hot interior. But the earth is a machine whose response to change can be to create the conditions which may threaten the agents of that change, and carbon dioxide provides a good example of that threat.

Human activity is both releasing carbon dioxide at an ever acceler-ating rate, and reducing the earth's ability to recycle it. As fast as we raise coal and oil to the surface and burn its carbon into carbon dioxide, so we also cut down the forests which contain the greatest concentra-tions of plants able to recycle that carbon dioxide. As a result, the carbon dioxide content of the atmosphere is steadily rising, and that is a problem to us because of the effect of that carbon dioxide on the escape of heat from the earth into space.

Some of the heat energy which is continually dispersing out into space in the form of infrared radiation can be absorbed by carbon dioxide molecules to increase their internal vibrations. So carbon dioxide serves as a chemical blanket insulating the earth against the loss of heat and so trapping that heat, initially within carbon dioxide molecules, but ultimately within the motions of all the particles in the atmosphere. So carbon dioxide is very effective at acting a bit like the glass of a greenhouse—letting the energy of sunlight flood into the earth while restricting the flow of radiant heat energy out. Carbon dioxide is certainly not the only gas in the air which absorbs infrared radiation in this way, but it is one of the most effective, and the one which human activity is increasing the most. The changes we are imposing on the carbon dioxide balance of the world have led to the well publicized fears of a global 'greenhouse effect', in which the rising levels of carbon dioxide (and other so called 'greenhouse gases') may raise the temperature of the atmosphere to an extent which might threaten our civilization and perhaps even our continued existence.

Nobody knows what the true effect of the undoubted rise in global carbon dioxide levels will be. Some predictions talk of melting polar ice-caps raising the sea level enough to flood most of the major cities of the world, and of the major crop-growing areas of the world being converted into hot and barren deserts. Some climatologists believe the first stages of such changes are already detectably under way, while others are not so sure; and a few suggest that the natural balance will soon restore itself (as plants increase their uptake of carbon dioxide, for example) so that no calamity need be feared. Other scientists point out that the earth should really have entered a new ice age by now, so perhaps the man-made greenhouse effect is all that is postponing that alternative catastrophe.

So what should we do? Carry on as normal and risk the disaster of melting ice-caps, flooded cities and prairies turned to desert? Or attempt to halt and then reverse the rise in carbon dioxide levels, perhaps only to remove our protection from the next ice-age? The issue is a very complex one which is not particularly well understood. It involves many factors other than simply carbon dioxide, and chemists and climatologists throughout the world are struggling to gain a better understanding of all the problems and possibilities so that any decisions we take are more likely to be good ones. Our ability to understand and predict the chemical changes in the atmosphere may determine the future of modern civilization.

We are busy altering the natural chemistry of the air in a vast number of other ways in addition to changing its levels of carbon dioxide, and many of these other alterations may also pose serious threats to our future health and survival. There are molecules in the air, for example, which are so rare that they account for much less than one out of every million of the air's particles, and yet everyone has heard of them and they may well determine how long you live, how soon you die. I am referring to the air's ozone.

Ozone is an unusual form of oxygen, in which three oxygen atoms are combined to form an O_3 molecule. It is formed when a two-step reaction is initiated by the absorption of some of the sun's ultra-violet rays by oxygen molecules. In the first step, oxygen molecules are split up—dissociated—by the input of ultraviolet energy:

$$+\text{u.v. energy}$$
$$O_2 \rightleftharpoons O+O$$

(double arrows indicate
the reaction is readily
reversible)

these oxygen atoms can then combine with oxygen molecules to form ozone:

$$O+O_2 \rightleftharpoons O_3$$

u.v. energy powers reverse
reaction of ozone
destruction

Ozone is an unstable molecule, because it too can be dissociated by the absorption of ultra violet radiation from the sun to regenerate normal oxygen molecules plus an oxygen atom, so the lifetime of any given ozone molecule is rather short.

The reactions that form ozone from O_2, and which reverse the process, combine to set up a constant dynamic equilibrium process high in the atmosphere. Various other reactions, in addition to the ones shown, contribute to the overall equilibrium levels of atmospheric ozone, and they all combine to maintain the earth's natural 'ozone layer', especially between the heights of 15 and 30 kilometres from the ground. The term ozone layer should not mislead you into thinking that within this layer ozone is very common. Even when at its highest concentration ozone is extremely rare, but the combined effect of the atmosphere's tiny concentrations of ozone on the sunlight that filters through the entire depth of the atmosphere is very significant indeed.

The ozone absorbs significant amounts of the ultraviolet radiation which is responsible for sunburn and tanning, and which can also initiate the damage to our DNA which causes cancer. So the ozone layer is important to us because it acts as a global sunscreen, reducing the strength of the penetrating ultraviolet rays to a level which most of us manage to live with. Even before human activity began to reduce the ozone layer above our heads, sufficient ultraviolet radiation penetrated through it to cause large numbers of cancers, especially of the skin. But in recent years many of the industrial chemicals we release into the atmosphere have begun to react with the ozone and so reduce its equilibrium concentration in the ozone layer. The resulting increased risks of sunburn and skin cancer have been well publicized.

A great many chemicals in the air, both natural and industrial, can react with ozone to diminish the ozone layer, but some of the most significant appear to be the 'chlorofluorocarbons' (CFCs) used by many industries and as the propellants in many aerosol cans. The effect of these chemicals is most significant not because they are the major contributors to the destruction of ozone, but because they are the major *additional* contributors, over and above the natural ozone destroyers. They have played a major role in shifting the natural balance between ozone formation and destruction in favour of destruction.

So once again we see that some of the rarest chemicals in air play vital roles in our lives, roles which we may dangerously interfere with by unnaturally altering the chemical balance of the atmosphere.

The atmosphere is a blanket of invisible chemistry which sustains and protects all life on earth. It contains the oxygen we must breathe to fuel the chemical 'fires' of life; it contains the carbon dioxide needed by plants to create the carbon-containing compounds which form much of the chemical structure of life, and which we must eat, directly or indirectly, in order to live and grow ourselves; it carries the water molecules which evaporate from the sea and are then deposited onto the land as the rain which we must drink or die; it contains the ozone molecules which protect us from some of the most damaging rays of the sun; it contains nitrogen molecules which our industrial chemists can incorporate into the ammonia and related chemicals we use as fertilizers; and of course it also contains all of the gaseous pollutants which we endlessly release into the air in the hope that they will be sufficiently diluted by the atmosphere that we can forget about them. Increasingly, however, our air pollution is returning to haunt us. Current concerns about the greenhouse effect and the ozone hole remind us that we should treat the complex chemistry of the air with great respect, and be very wary of interfering with it.

10　Water

Water is one of the simplest chemicals of the earth, but one of the most important and most abundant. The view from space suggests that 'water' would be a more appropriate name for our planet than 'earth'. We live on the water planet, most of the surface of which is covered in deep oceans and seas of that simple liquid. The water lying on the earth's surface rises up into the atmosphere as water vapour, and then falls back to the surface as the rain which washes and shapes the land and brings into the sea the chemicals needed to form new sedimentary rocks. All the living things which flourish upon the land are largely composed of water—the chemistry of life occurs within watery internal seas, each forming the 'cytosol' of one of our trillions of cells. Water is the 'solvent of life', the liquid, in other words, in which most of life's chemistry takes place, and without which that chemistry would be impossible.

All of the chemical activities of water must be dependent on some fundamental aspects of its simple molecular structure, consisting of a single oxygen atom bonded to two hydrogen atoms. In fact, just one simple fact about the water molecule is crucial—its oxygen atom exerts a very strong pull on the shared electrons of the molecule, while the hydrogen atoms exert a much weaker pull (for reasons discussed in box 6.1). This difference in electron-pulling power of oxygen and hydrogen makes the two bonds of a water molecule strongly polarized, creating a significant region of partial negative charge around the oxygen atoms, and significant regions of partial positive charge around the hydrogen atoms. Most of the important chemical activities of water are a result of the strong polarization of the oxygen to hydrogen bond.

The polar covalent nature of water molecules allows a great variety of ionic substances to become dissolved in water. By becoming 'dissolved', we mean that the ions of an ionic substance become separated from one another and evenly intermingled with the water molecules. For this to happen, the positive and negative ions must be able to fit into

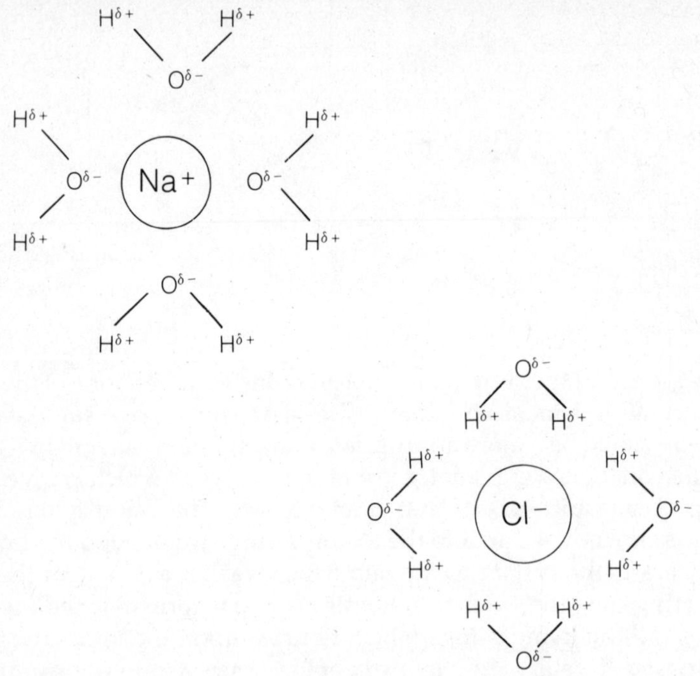

Figure 10.1   Water molecules surround ions in a way which allows them to become dissolved in the water.

the water structure in a way which prevents them from being pulled together by the electric attraction between them. What happens is that each ion becomes surrounded by water molecules as shown in the example of figure 10.1. The water molecules are pulled, by the electric force, into configurations with their δ^+ regions facing the dissolved ion if it is a negatively charged ion, and their δ^- regions facing the ion if it is a positive ion. In this way a 'cage' of water molecules around each ion tends to insulate the ion from other oppositely charged ions, allowing the ions to stay apart and remain dissolved in the water rather than being pulled together into a solid ionic precipitate.

So the polar covalent nature of the water molecule allows ionic compounds to dissolve in water. It made the sea 'salty', by allowing it to carry dissolved ions such as sodium (Na^+) and chloride (Cl^-) ions, and it allows the water of your blood and plasma and cell cytosol to carry the complex balance of ionic substances which makes the chemistry of life possible.

Lots of non-ionic compounds dissolve in water as well, but they all tend to consist of molecules containing some polar covalent bonds. These bonds generate δ^+ and δ^- regions on the molecules which can interact with the partial charges of the opposite sign on water molecules and so allow the dissolved molecules to slip easily into the water structure. So the ability of water to act as a solvent for such molecules again depends on water's polar covalent bonds. These bonds also explain why oily, fatty substances will not dissolve in water. If you examine the chemical structure of the substances we call fats or oils, you find that they consist of molecules, usually rather large molecules, held together by bonds which are not strongly polar covalent. So these molecules do not carry the δ^+ and δ^- regions required to interact with water molecules and so allow them to dissolve in water. Molecules like this tend to cluster together instead, to form oily globules in which the molecules can interact via Van der Waals bonds between themselves.

So the polar covalent nature of the water molecule makes water a discriminating solvent, able to dissolve many ionic and polar covalent substances but not 'non-polar' fatty substances. In later chapters you will see how water's discriminating powers as a solvent give it great influence over the structure and activities of the proteins and membranes of living things, and even such mundane matters as the ability of soaps and detergents to wash your clothes. Without the discriminating powers of water as a solvent the world would be a very different place, and we would not be here to observe it.

There is a lot more to the chemistry of water, however, than its powers as a solvent. Water is a direct participant in many chemical reactions, far too many for us to mention them all here. What we can do is briefly examine the central role water plays in two very common and important types of reaction, in which many small molecules become linked up into chains, or, conversely, long chain like molecules become broken up into their smaller constituents.

Figure 10.2 illustrates the essential chemical principles behind a great many 'polymerization' reactions, in which small molecules unite to form large 'polymers' with the accompanying formation of water. The water comes from an O–H ('hydroxyl') group protruding from the end of one molecule, and a hydrogen atom protruding from the end of another. In chemicals which undergo this sort of polymerization the oxygen atom carries a δ^- charge, because it is bonded to a considerably less electronegative atom, while the lone hydrogen on the other molecule carries a δ^+ charge because it is bonded to a more electronegative atom. These polarizations of the two bonds encourage them to break, under appropriate conditions, in a process which can be regarded overall as the removal of a fully negatively charged OH^- group and a

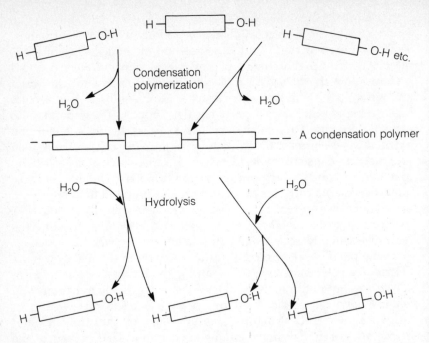

Figure 10.2 Condensation polymerization can create long polymers from many much smaller monomers. Hydrolysis can break up the polymers back into their monomers.

positively charged hydrogen ion (H^+) from the respective chemicals, followed by the union of the H^+ and the OH^- into a water molecule and the linking together of the rest of the two original molecules by a new bond. Further suitable molecules can then add onto both ends of this growing chain via the same chemical reaction, which can proceed until a very long chain has formed and many water molecules have been released.

Polymerization reactions like this are known as 'condensation' polymerizations, due to the formation of water as they proceed. Condensation polymerizations serve to create many natural and unnatural polymers from their smaller 'monomer' building blocks. These monomer units may be identical molecules, or quite different molecules—the variety of condensation polymerizations is enormous.

Of course the polymerization reaction can also be reversed, given appropriate conditions and perhaps catalysts, allowing water molecules to 'add across' the bonds holding the monomers together to recreate the free monomers. This is known as the 'hydrolysis' of a polymer, hydrolysis being the chemical term for the chemical decomposition of a substance due to its reaction with water.

In any discussion of chemistry it is usually not long before water appears in some role or another. We have already seen how water is formed as one of the products of combustion, and how it is used by plants as one of the raw materials for the manufacture of sugars. There is water stored as vapour in the air around us, condensed into liquid form in lakes and rivers and the sea, and locked up into the solid structure of ice towards the colder poles of the planet. Water is the solvent in which the chemistry of life occurs, and is also used by chemists as a solvent for many other types of chemical reactions. It controls the structure and activities of many of the chemicals of life; and it actually participates directly in a virtually infinite variety of chemical reactions, either being formed from oxygen or hydrogen, or fragmented in some way, depending on the particular reaction concerned. The tiny molecule H_2O contains within its almost ridiculously simple structure an astonishing chemical potential, largely due to the particularly unequal sharing of electrons within its two identical polar covalent bonds.

11   Carbon

Of all the substances we mine from the earth few can be more cherished than diamond. The fascination and beauty of diamonds makes us willing to pay astonishing prices simply to be able to admire them as they sparkle in a bracelet or a ring, or as they dangle from an earlobe. Industry values diamond highly too, for it is the hardest substance we know. It can be used to make extremely tough cutting tools and is exploited in many other situations where toughness and durability are at a premium. At the other end of the value scale of earthly materials, few substances can be less treasured than the soot and charcoal left behind when many fuels are incompletely burned. Although soot and charcoal do have their uses, to most of us they are messy irritations. For diamond and soot, read riches and rubbish, yet chemically they are virtually identical. In their pure forms both are composed solely of carbon atoms. The only difference between diamonds and soot is the way in which their carbon atoms are linked together by chemical bonds. What great value chemical bonds can have!

Chemical bonds involving carbon atoms are some of the most significant on the earth. Not only do they make the difference between diamond and soot, they also hold together most of the chemical structure of life and make the chemistry of life possible. This all makes them worthy of some further exploration.

A carbon atom consists of a nucleus containing six protons, usually six neutrons, and surrounded by six electrons. The principles of atomic structure outlined in chapter 4 would lead us to expect the electrons of carbon atoms to occupy orbitals as outlined in the top of figure 11.1, with one filled and two half-filled orbitals in the outer shell. This picture causes problems when we come to consider the actual chemical reactions of carbon, and leads us towards a new and important concept of bonding theory. Carbon atoms tend to react to form four bonds with other atoms, such as the four single bonds of a methane molecule (figure 11.2a). Diamond consists of an extended network of carbon

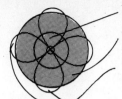

Two electrons in inner spherical 's' orbital

Two electrons in outer spherical 's' orbital

One electron in each of two dumb-bell shaped 'p'
orbitals of the outer shell, with one other
'p' orbital (not shown) remaining empty

Simple atomic theory suggests the electron arrangement of carbon atoms
should be as shown above. However, when carbon atoms react they
often behave as if their four outer orbitals – one spherical 's'
orbital and three dumb-bell shaped 'p' orbitals – become mixed or
'hybridized' to generate four identical 'sp³' hybrid orbitals,
as shown below

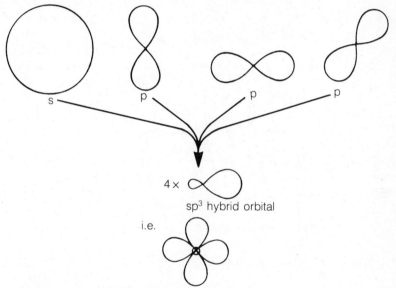

Figure 11.1 Hybridization of the four outer orbitals of carbon atoms.

atoms each linked to four other carbon atoms by four identical bonds
(figure 11.2(b)). At first sight, the valence bond theory of chemical
bonding seems to suggest this should not happen. Remember that this
theory views covalent bonds as arising when half-filled orbitals on one
atom overlap with half-filled orbitals on other atoms. A carbon atom as
described above has only two half-filled orbitals available, so we should
expect it to form only two bonds. How can each carbon atom of a
diamond be bonded to four other carbon atoms, or each carbon atom of
a methane molecule be bonded to four hydrogen atoms? The answer
introduces the new concept of orbital 'hybridization'.

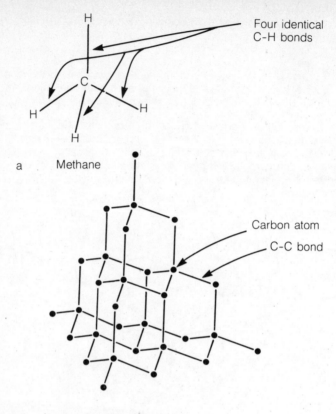

a   Methane

Carbon atom

C-C bond

b   Diamond (showing only a tiny
portion of the structure of a diamond)

Figure 11.2   The structure of methane and diamond.

It turns out that the basic set of orbitals available for electrons, listed in figure 4.3, are not the only possible orbitals. Alternative 'hybrid' orbitals exist which can be regarded as being the product of mixing, or hybridizing, any number of the basic orbitals. In the outer shell of a carbon atom, for example, we have four basic orbitals available: a spherical one designated as an 's' orbital, and three dumb-bell shaped 'p' orbitals. The p orbitals are of higher energy than the s orbital, which is why we originally assigned two of carbon's four outer shell electrons to the s orbital, and the other two to two of the p orbitals with one p orbital remaining unoccupied. Nature is not as simple as this, however. The one s and three p orbitals can be replaced by four hybrid orbitals, each a little like an s orbital and a lot like a p orbital. The shape of these

hybrid orbitals is shown at the bottom of figure 11.1. The most important point about them is that they are identical in both shape and
energy. So if we imagine the one s and three p orbitals of the outer shell
of a carbon atom combining to produce these four identical hybrid
orbitals, and then imagine one of the four outer electrons occupying
each orbital, we attain a form of carbon which has four half-filled
orbitals available to overlap with other half-filled orbitals on other
atoms to form four covalent bonds.

This is the way in which chemical theory rationalizes the structure of
methane, diamond and millions of other carbon-containing compounds,
and the fact that carbon atoms very often behave as if they possess four
identical half-filled orbitals rather than two. It might seem like a fiddle
contrived just to make things fit, and in a sense it is, but it is a fiddle which
seems to give us a very good description of the way the chemical world
really behaves. The electrons around atoms really do behave as if they can
occupy a whole series of hybrid orbitals as well as the most basic and
'pure' orbitals predicted by the wave equation. Of course this opens up a
whole new range of possibilities when we consider the precise nature of
the orbitals around atoms and within molecules.

One further variety can be illustrated by looking at the structure of
another form of carbon known as graphite. In graphite, the carbon
atoms appear to be ordered into flat sheets of atoms with each carbon
atom bonded to only three others (see figure 11.3). This alternative
structure for an assembly of carbon atoms to adopt is rationalized using
a different form of hybridization. It is supposed that the outer s orbital
and only two, rather than all three, of the outer p orbitals hybridize
together. This yields three hybrid orbitals, as shown in the figure, and
one remaining unhybridized p orbital. Puting an electron in each of
these four orbitals (remember carbon atoms have four electrons in their
outer shell) yields three identical half-filled hybrid orbitals, which are
the ones which overlap to form the covalent bonds holding the graphite
sheets together. Each carbon atom also carries a lone electron in a p
orbital protruding above and below the plane of bonded atoms.

Graphite, unlike diamond, is a good conductor of electricity, and it is
the lone electrons of the outer p orbitals which are believed to be able to
flow through the graphite in order to allow an electric current to pass.
The structure of graphite leaves these electrons free to 'hop' from atom
to atom, and thus be part of an electric current passing through the
graphite, because they are not involved in bonding the carbon atoms
together. Graphite is also physically a very different substance from
diamond. It is easily broken, often into flattish plates, and it can act as a
solid lubricant or as the 'lead' of pencils. These properties all seem to
stem from the fact that the individual sheets of atoms in the graphite

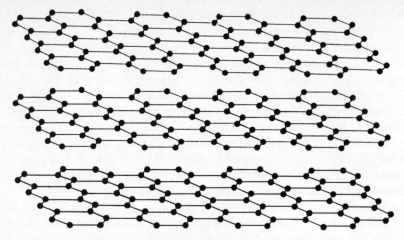

Graphite consists of stacked sheets of carbon atoms, each atom bonded to three other atoms (dots represent atoms, lines represent covalent bonds)

This structure is explained by assuming that the spherical outer 's' orbital of each atom becomes hybridized with two of the outer 'p' orbitals. This partial orbital hybridization yields three identical 'sp^2' hybrid orbitals and one unhybridized p orbital as shown below. Each orbital contains one electron, but only the sp^2 hybrid orbitals overlap with the similar orbitals of neighbouring atoms to form covalent bonds

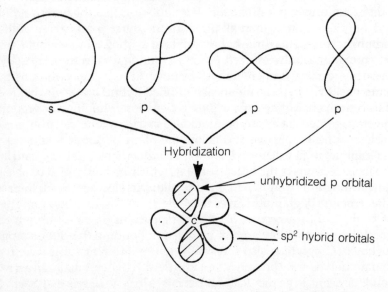

Figure 11.3　The structure of graphite and the theory explaining it.

structure are not covalently bonded to one another. This allows them to slip over one another easily and shear away from one another, allowing the graphite structure to be easily broken, allowing a thin layer of graphite to slip from a pencil and onto paper as you write, and allowing the 'slippery' graphite sheets to form a solid lubricant.

The theory of orbital hybridization allows us to offer a very convincing explanation for why diamond is extremely hard and will not conduct electricity, while graphite is soft and slippery and a good conductor. I have looked at these structures from the point of view of valence bond theory, since that is the way chemists normally think about them, but equally good descriptions are available using molecular orbital theory.

A third form of carbon is known as amorphous carbon, in which there are no regions of extended ordered structure corresponding to either the diamond or graphite form. Soot is largely amorphous carbon, in which there may be many tiny regions of diamond-like and graphite-like structure all mixed up with more irregular arrays of linked atoms. It is tantalizing to think that the soot from a fire may contain millions of tiny sub-microscopic diamonds!

So the differences between the messy rubbish of soot, the slippery softness of graphite and the hard brilliance of diamond depend solely on the variety of electronic structures and bonding possibilities available to carbon atoms when they interact.

Some people might think that the ability of carbon atoms to form diamonds is sufficient reason for us to be interested in their structure and chemical activities. There is much more richness to the chemistry of carbon, however, than the creation of mere diamonds. As I said earlier, the chemistry of carbon lies at the heart of the greatest treasure found on earth—the treasure of life. In fact, the chemistry of carbon containing compounds has become known as 'organic' chemistry, in recognition of its central importance to the chemistry of living organisms. You see, there is something rather special about carbon atoms. First, they can form strong covalent bonds with up to four other atoms, giving them the ability to link together into large and very varied molecules in which chains of bonded carbon atoms are bonded to various other atoms and chemical groups as well—either attached as 'side-groups' of the carbon chains or incorporated as bridging links between the carbon atoms. Secondly, the electronegativity of carbon atoms lies pretty much in the middle of the range of electronegativity values, allowing them to form covalent bonds with a very wide range of different types of atoms, but often rather strongly *polarized* covalent bonds (i.e. polar covalent bonds) which incorporate the δ^+ and δ^- charges which are so powerful at mediating many subtle and varied chemical effects.

When you look at the structure of the chemicals that make life possible, and which control all of its activities, you are immediately struck by the predominance of carbon atoms in central roles. It is time for us to examine the most significant of these chemicals of life, an important and fascinating part of our investigation of chemistry which will occupy us for the next four chapters.

12   Genes

The chemical known as DNA, which stands for 'Deoxyribonucleic acid', is the stuff of our genes, the stuff of heredity. It is a chemical sculpture—a structure of almost mystical beauty and symmetry which decorates the entrance halls of many laboratories of chemistry and biology. A model of DNA is an attractive object even to someone with no knowledge of its significance. Coloured spheres of appropriate size represent the atoms, which may merge into one another in a realistic representation of DNA, or may be connected by thin metal rods representing the chemical bonds that hold the structure together. As you look at a model you can quickly identify two helices of linked atoms winding around a central core—the helices which make the DNA structure a 'double helix' (see figure 12.1).

DNA is the chemical which determines what sort of body we have, what we look like, what intellectual potential we have, what diseases we are most likely to suffer from, what species we are. The differences between you, me, a baboon, a blackbird, a blackcurrant bush and a bacterium are believed to be determined by the chemical structure of our DNA. What magical chemistry gives DNA such power over the shape of life? In fact, of course, there is no magic, for the power and activities of DNA can be explained using the simple fundamental principles of chemistry outlined in the first seven chapters of this book. DNA is a chemical and its activities are chemical activities which depend on the pushing and pulling of the electromagnetic force and the energy changes which accompany the dispersal of energy towards a more even distribution.

DNA is a dead, powerless and helpless molecule on its own. To be of any significance it must interact with the other chemicals around it within a living cell. Each of these other chemicals is also dead and helpless on its own, for they too must interact with the chemicals all around them. The phenomenon of life is the net effect of a complex network of chemical reactions all influencing and being influenced by

129

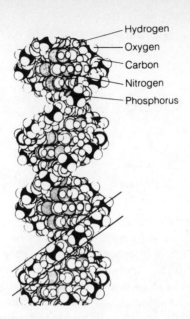

Hydrogen
Oxygen
Carbon
Nitrogen
Phosphorus

Figure 12.1   The structure of the DNA double-helix. The 'molecular back-bone' of part of one helix is highlighted by the lines drawn on either side of it.

one another. Of all these chemicals, however, DNA has a particularly vital role to play in the living machine. First, its structure largely determines what chemical reactions proceed within a living thing, because it determines the particular suite of biochemical catalysts each living cell contains. The catalysts themselves are called proteins, and we will be examining them in chapter 13, but although these proteins directly determine the chemical nature of a living cell, the cell's DNA is the ultimate determinant because it determines what proteins a cell contains. Secondly, DNA molecules have the unique ability to undergo 'replication'. They have the ability, in other words, to give rise to two copies of themselves where previously there was only one. This is a kind of molecular reproduction which lies at the heart of the reproduction of all forms of life. Bacteria can multiply into a life-threatening infection, weeds can spread across an untended garden, and you can generate children, all because DNA has the ability to replicate. The replication of DNA is what allows life to 'go forth and multiply'.

So DNA is vital to living things because it can, indirectly, determine what chemical reactions proceed within living things, and because it can give rise to copies of itself and so allow living things to reproduce. We must look at how these two great feats are achieved by a 'mere molecule'.

DNA is composed of only five different types of atom—hydrogen, carbon, nitrogen, oxygen and phosphorus. These atoms are bonded together into the DNA molecule, although strictly speaking each DNA double helix consists of two DNA molecules wound around one another, as we shall see. Also, strictly speaking, DNA is not a molecule but a 'molecular ion', since it contains some oxygen atoms which have been ionized to carry a negative charge, as you can see from figure 12.2.

Within the structure of a long double helix we can identify a rather simple repeating pattern and some simplifying structural characteristics. Each individual helix can be thought of as possessing a molecular 'backbone' consisting of a group of atoms known as a sugar group, connected to a group known as a phosphate group, connected to another identical sugar group, then another phosphate group, and so on. The helix, in other words, contains a 'sugar-phosphate' backbone. The chemical groups which form the central core of a double helix are linked to the sugar-phosphate backbone via the sugar groups, and there is simplicity to be found here as well. Each sugar group is bonded to one of four chemical groups known as bases. The names of the four bases are adenine, thymine, guanine and cytosine, otherwise known as A, T, G and C. If you examined the structure of a double helix further you would soon discover why I said it consists of two molecules wound around one another, and not one. Each base protrudes into the centre of the double helix to be held close to another base carried by the opposite DNA strand. The two opposing bases are not held together by full covalent or ionic bonds, however, but are held together (and so the entire double helix is held together as a double helix) by weaker forces of attraction between the δ^+ and δ^- charges on atoms involved in polar covalent bonds. You can see these bonds in figure 12.2, and since hydrogen atoms with δ^+ charges are involved in all of them they are all 'hydrogen bonds' of the type discussed on page 70.

So the DNA double helix consists of two strands of the DNA molecular ion, wound around one another and held together by the force of attraction between atoms carrying partial electrical charges because they participate in polar covalent bonds.

The structure of the bases ensures that the base A can only form a 'base pair' with the base T, and vice versa, while the base G can only pair up with C, and vice versa. So only two base pairs are possible, A-T and G-C, although they can appear either way round. This chemical 'complementarity' between pairs of bases of DNA is what allows a double helix to undergo replication.

During the process of DNA replication, shortly before a cell divides in two, for example, the two complementary strands of the double helix

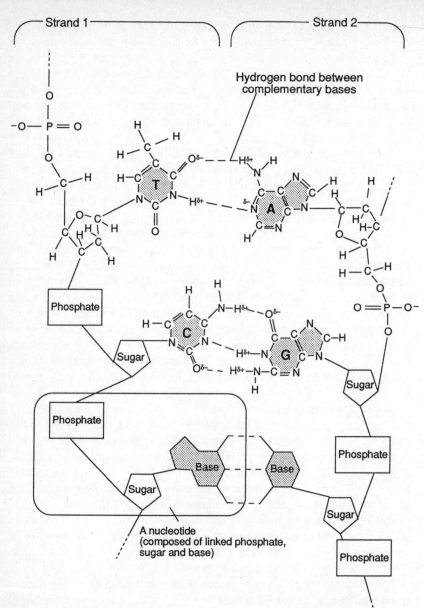

Figure 12.2   The detailed structure of DNA, with the strands unwound for clarity. The top base-pair (A–T) and its linked sugar and phosphate groups are shown in full. The other type of base-pair (G–C) is also shown in full in the middle of the diagram; but all other bases, sugar groups and phosphate groups are shown in schematic form. One 'nucleotide' unit is outlined at bottom left. Note that the two strands of DNA each consist of a series of nucleotide units linked together; and that the two strands are antiparallel (i.e. they run in opposite directions).

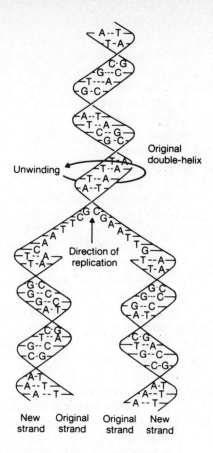

Figure 12.3   The replication of double-helical DNA.

unwind. This makes each strand available to serve as a kind of molecular template or mould on which a new complementary strand can form (see figure 12.3). A new strand forms when chemicals known as nucleoside triphosphates, each carrying a base, a sugar group and three phosphate groups, become bound to the pre-existing DNA strand and then undergo the reaction which links the sugar, the base and one of the phosphate groups into a new strand of DNA. The nucleoside triphosphates become bound thanks to the ability of each of their bases to bind by base-pairing to one particular base in the existing DNA. So the simple chemical process of base-pairing, which holds the double helix together, is also responsible for allowing one double helix to generate two 'daughter' double helices.

The actual linkage reaction, in which a newly arrived base, sugar and phosphate group becomes linked into a growing DNA strand, is catalysed by the action of the biological catalysts known as enzymes (see chapter 13), but the *specificity* of the reaction, the fact that only the 'correct' bases are incorporated at each point, depends on the structure of the bases and the fact that each one can form a suitable base pair with only one other type of base. So some rather simple principles of chemical structure and the weak attraction between appropriately positioned δ^+ and δ^- charges allow the double helix endlessly to regenerate the genetic information needed to specify the structure and activities of all life.

What form does the genetic information stored within DNA take, and how does it hold such control over living things? Remember that I told you that the real importance of DNA is that it determines the particular suite of biochemical catalysts each living cell contains. These catalysts are all protein molecules, and each specific protein consists of a sequence of individual 'amino acids' linked into a long protein chain. The structure of amino acids does not matter for the moment. The essential point is that to construct any specific protein molecule a specific sequence of amino acid molecules must become linked together into a chain. The influence of DNA over the chemistry of life is due to the ability of specific sections of a double helix to determine the amino acid sequence of specific proteins.

The process by which the genetic information stored in DNA gives rise to functional protein molecules involves many steps and many distinct chemical reactions, all catalysed by appropriate enzymes which have previously been made in the same way. It is easy, however, quickly to outline the main features of the process and the essential chemical simplicity that lies behind it.

A particular section of any double helix which specifies the structure of any one protein molecule is called a 'gene', and the conversion of the genetic information held in the gene's structure into the form of a protein molecule is known as 'gene expression'. The first step in the expression of a gene is a form of replication, except only one strand of the gene is used as a molecular template, and the new molecule formed on it is not DNA, but a very similar chemical known as RNA. There are only two differences between RNA and DNA. Each sugar group of RNA carries an extra oxygen atom, and a base known as uracil is used in RNA in place of the very similar base thymine found in DNA. Uracil behaves just like thymine when forming base pairs, so it pairs up with adenine alone.

Once the single stranded RNA copy of one DNA strand of a gene has been made, it is transported to another site in the cell where it becomes bound to large complexes of protein and RNA known as ribosomes, and it is at these ribosomes that the RNA, known as 'messenger RNA' (mRNA) specifies the structure of a protein molecule. It does so thanks to another kind of RNA, known as 'transfer RNA' (tRNA), which acts as the chemical intermediary between the structure of the mRNA and the structure of the protein it gives rise to. Twenty different amino acids are used to construct proteins, and for each amino acid there is at least one specific tRNA which can react with the amino acid so that it carries it attached to one end of the large tRNA molecule. At the other end of each tRNA, there is a group of three bases known as an 'anticodon'. These three bases are able to bind to three complementary bases of mRNA, known as a 'codon', when they are exposed at a special site on the ribosome. Thanks to the formation of base pairs between the bases of mRNA codons and tRNA anticodons, the appearance of a specific codon in mRNA results in a specific amino acid being brought to the ribosome. The ribosome and the mRNA actually slide along one another, so that, one by one, all of the codons on the mRNA have the opportunity to bind to appropriate tRNAs which each brings with it a specific amino acid; and as all the different amino acids arrive at the ribosome, a chemical reaction links them up into the growing protein chain (see figure 12. 4).

As I am sure you appreciate, I am summarizing the most basic details of an incredibly complex chemical process, and each step of the process is catalysed by the activity of one or more specific enzymes; but it is just a chemical process.

The chemical structure of the DNA of a gene causes a protein with a specific chemical structure to be formed simply due to the interactions and reactions of a great many chemicals within the complex chemical 'ocean' of the living cell.

The essential link between the structure of a gene and the structure of a protein is that a particular sequence of bases in the gene corresponds directly to a particular sequence of amino acids in the protein. A gene consists, essentially, of a sequence of bases strung out along a sugar-phosphate backbone; while a protein consists of a sequence of amino acids joined end to end. As we move along a gene we find that each group of three bases corresponds to one specific amino acid in the protein that the gene 'codes for'. So that is the secret of the celebrated 'genetic code', and it is the essential chemical secret which allows the DNA of our genes to have such a powerful influence over the structure and the activities of our bodies.

We have looked at the structure of the DNA of our genes; and examined in outline how that structure allows genes to be replicated, to generate the new copies that must be handed on to future generations,

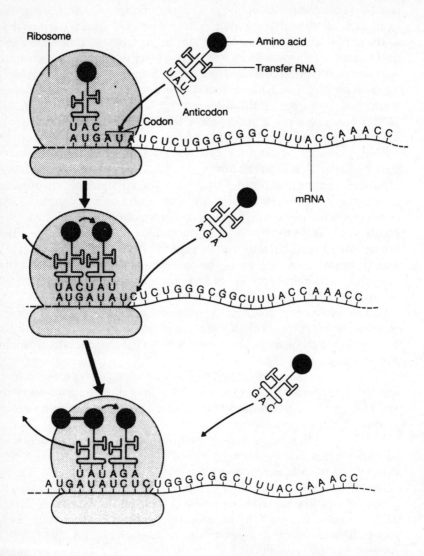

Figure 12.4   The translation of genetic information into proteins. A ribosome moves along mRNA, allowing transfer RNAs to bind to successive codons and let the amino acids carried by the tRNAs be linked together into a protein chain. Linkage of the first two amino acids of a protein is shown. Repetition of this procedure allows a long mRNA to direct the manufacture of a long protein chain.

and at how it allows genes to specify what protein molecules a living thing contains. If that is all that genes do, then the proteins must be the real molecular 'workers' that actually do all the chemical work of constructing living things from lifeless raw materials. This is indeed the case, so it is time for us to take a look at proteins to try to discover the chemical principles behind their seemingly magical powers.

13   Proteins

Proteins are large, complex and truly remarkable chemicals. You are a chemical machine built and maintained by the activities of proteins. Protein molecules form much of the structural framework of your cells, bones, muscles, connective tissue and skin. Almost all of the chemical reactions involved in allowing you to live and grow and move about and think are catalysed by the special class of proteins we call enzymes. Proteins in your red blood cells transport oxygen from your lungs to every cell of the body, bringing supplies of that vital gas to the food molecules which must slowly 'burn' in it to provide you with energy. Proteins act as some of the most vital 'hormones' (such as insulin) which are released by some cells of the body to travel around the bloodstream and control the chemical activity of other cells. Proteins can bind to the DNA of your cells to switch particular genes on and off when appropriate, and they can also bind to other proteins to modify their activity as well. The antibodies which latch on to invading micro-organisms and help to neutralize them, and which save us all from certain death many times in our lives, are protein molecules; and proteins do much more besides all that. They are the chemical masters of life, even though the 'information' required to generate them is stored in the structure of the DNA of our genes.

In chapter 12 you saw how the base sequence of a gene is able to determine the structure of a protein by determining which amino acids are brought to the ribosome to be linked into a protein and the sequence in which they should be linked. Now it is time to look at what sort of chemicals these amino acids are, how they become linked together and how they allow proteins to perform their astonishing range of chemical tasks.

The way in which 20 different amino acids serve as the basic building blocks of proteins is shown in figure 13.1. There is a welcome structural simplicity about the chemistry of all 20 protein-making amino acids: they all share an identical 'backbone', linked to one of 20 possible 'side

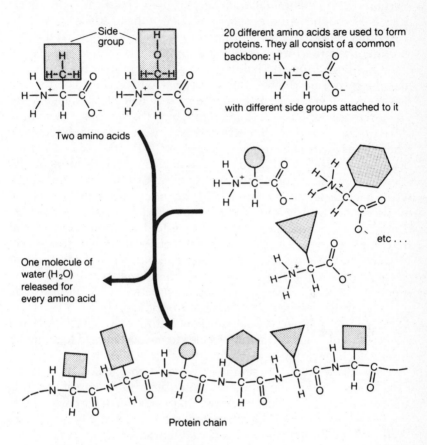

Figure 13.1 Proteins are composed of long chains of linked amino acids. Twenty different types of amino acids are found in proteins, and most proteins are hundreds of amino acids long.

groups'. The atoms of the backbone are the ones that link up to hold a protein chain together, while the side groups make each protein unique and give it the chemical power to do whatever it is that it does. Figure 13.1 shows you the reaction which allows amino acids to become linked into a protein chain. This is essentially a simple 'condensation' reaction of the type considered in chapter 10. At the end of the reaction two hydrogen atoms and an oxygen atom from the amino acids have become united into a molecule of water, leaving the two amino acids linked by a bond between a carbon and a nitrogen atom. This is the reaction which joins amino acids into a protein chain when the amino acids are brought to the ribosome by their transfer RNAs.

Another welcome and vital simplicity of the proteins is that the task of protein manufacture is essentially complete as soon as the protein chain has been formed. That is not to say that protein molecules are long floppy chains; they are not. For a protein to be able to perform its task, as an enzyme, a structural protein, or whatever, it must fold up into a rather specific conformation in which the various atoms of the side groups and the backbone are brought into a particular orientation which gives the protein its chemical power. This vital folding process, however, proceeds *spontaneously* as soon as a protein is formed. The specific amino acid sequence of a newly formed protein, and the way in which it interacts both with itself and with the chemicals in the environment, spontaneously forces the protein into its final active folded conformation. So, in other words, proteins fold up automatically due to the chemical interactions between the various atoms of the protein and between the protein's atoms and the chemicals, especially water molecules, surrounding the protein. The most important influence on protein folding is probably the interaction between the atoms of the protein and the molecules of water surrounding it.

We have seen how the polar covalent nature of the two bonds in a water molecule causes water molecules to be loosely held together by an extended network of electromagnetic attractions between the δ^+ charges on the hydrogen atoms and the δ^- charges on the oxygen atoms. Some of the side groups of amino acids have a great chemical affinity for water, essentially because they carry full or partial electrical charges which can easily be integrated into the network of interacting charges between the water molecules themselves. Such amino acids are called 'hydrophilic' (water loving). The side groups of other amino acids have no chemical affinity for water at all, essentially because they carry no regions of full or partial electrical charges, and so they are unable to integrate into the network of interacting charges in any body of water. Such 'hydrophobic' (water hating) amino acids experience a force pushing them together away from the water molecules and into a tightly bundled 'core' in the interior of a folded protein. The origin of this force is really the force of attraction between water molecules, since this tends to 'squeeze' the hydrophobic amino acids out of the water structure as much as is possible. Another way of putting it is to point out that the bundling of hydrophobic amino acids together in the core of a protein lowers the energy of the whole system, since it takes energy to disrupt the network of electric attractions between water molecules by pushing a hydrophobic amino acid into the middle of it. In the normal circumstances prevailing in a living cell, the dispersal of energy away from a protein will ensure it settles down into a rather low energy conformation with the hydrophobic amino acids hidden away from the surrounding water as much as is possible, given the amino acid sequence of the protein, and with the hydrophilic amino acids interacting with the water as much as the amino acid sequence allows.

Once hydrophobic amino acids have been pushed together in this way, they will themseves be held together by the weak forces of electromagnetic attraction known as Van der Waals bonds.

Other forces are also involved in making proteins fold into a specific conformation. Various amino acids carry full positive or negative charges, and many others carry the partial charges generated by polar covalent bonds. Oppositely charged regions of a protein can be drawn towards one another, and similarly charged regions forced apart. Finally, some parts of a protein chain can become 'cross-linked' to one another by the formation of full covalent bonds between two sulphur atoms. Such 'disulphide bridges' tend to hold the final folded conformation together, giving it added stability and rigidity, rather than bringing about that conformation in the first place.

So protein chains fold up spontaneously into a rather specific conformation, thanks, essentially, to the pushing and pulling of the electromagnetic force and the dispersal of energy towards a more even distribution—thanks, in other words, to the same factors that bring about all chemical change.

These same factors, of course, must be responsible for all of the things that proteins can do. The essential secret of protein activity is that the folding of different protein chains can generate a great variety of chemical surfaces on which the electromagnetic force can push and pull at the electrons of the chemicals the proteins interact and react with. The shape of that chemical surface and the particular atoms and ions present on it are the two crucial features determining what any protein can do.

The atoms and ions present on a protein are those of the repeating amino acid 'backbone', plus those of the amino acid side groups. Some of the amino acid side groups have full negative charges, some have full positive charges, while some have no areas of strong charge at all. There are small side groups, and bulky side groups, and all sizes of side groups in between. There are atoms which carry a δ^+ charge (such as any hydrogen atom attached to an oxygen atom), and there are others which carry a δ^- charge (such as all of the oxygen atoms connected by double bonds to carbon atoms). There are long chains of atoms, branched chains, and a few instances of atoms joined together into rings. All in all the suite of 20 amino acids available for the construction of proteins contains chemical groups which, once brought into appropriate contact with one another, can perform just about any chemical task we could think of.

The chemical power of proteins, however, is not dependent on only the amino acids of the main protein chain. Many proteins fold up to create sites to which other chemicals can become bound, creating a hybrid molecule whose power depends on the combination of a protein with something else. The activity of many proteins, for example, depends on their ability to bind to small ions such as Zn^{2+}, Ca^{2+}, Cl^-, etc. These become bound to the proteins due to the electromagnetic attraction between them and full or partial electric charges on the protein. The activity of many proteins also depends on their ability to bind to fairly large molecules or molecular ions known as coenzymes. These form a rather loose association with the protein, altering the chemical nature of its surface in ways which allow the protein to do things which it would otherwise be unable to do. In some cases non-protein chemical groups become permanently attached to protein molecules, due to the formation of covalent bonds between the proteins and these modifying groups.

So the eventual chemical activities of proteins are not all produced solely by the atoms belonging to the proteins' amino acids. The atoms and ions belonging to other chemicals which become bound to the proteins can also have roles to play, but these other chemicals are able to bind to the proteins because the protein structure allows them to do so, by forming appropriate binding sites; so the prime importance of the folded structure of a protein to its eventual activities remains intact as a fundamental principle, despite the complications just summarized.

Enough of general principles: it is time to take a look at one particular protein in action, to gain an insight into the sorts of chemical mechanisms which allow all proteins to perform their actions.

At this moment enzymes in your gut are probably reacting with some of the protein molecules in your last meal. We need protein in our food as one of the vital raw materials for the growth and maintenance of our bodies. That protein is broken up into amino acids as our food is digested, and these amino acids are then either used directly as the raw materials for building new proteins, or else degraded further into simpler chemical supplies. One of the enzymes which splits proteins up into smaller pieces is known as chymotrypsin. Its job is to catalyse the reaction shown in figure 13.2, in which the bond holding two amino acids together is broken ('hydrolysed') during a reaction with water.

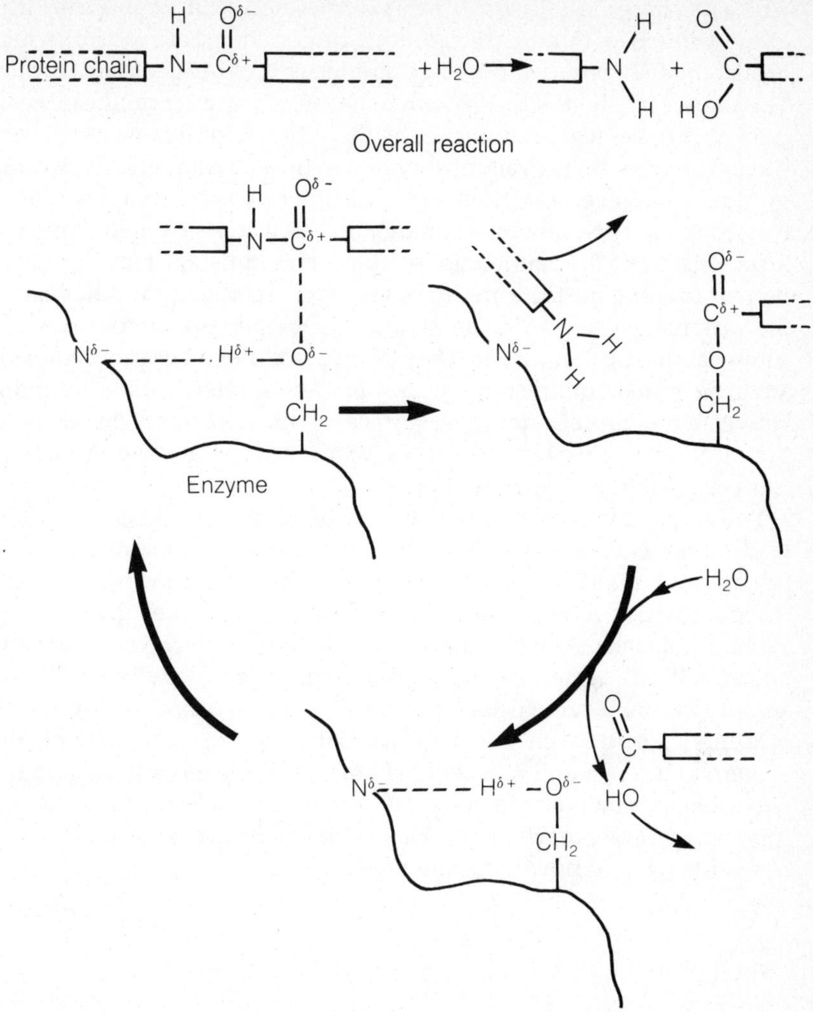

Figure 13.2   Some details of the mechanism of action of the enzyme chymotrypsin.

This reaction is one of those which releases energy as it proceeds, but which involves a rather high activation energy. It is very unlikely to proceed spontaneously without the catalytic assistance of the enzyme.

The enzyme achieves its catalysis in a series of steps, only the most vital of which are shown in figure 13.2. An oxygen atom on the enzyme with a δ^- charge is attracted towards the carbon atom of an amino acid of the protein with a δ^+ charge, initiating a chemical reaction which breaks the C–N bond holding two amino acids of the protein together. This allows one half of the protein to be jostled away from the enzyme (probably to be degraded further by the action of other enzymes), and leaves the other half covalently bonded to the enzyme surface. A water molecule, however, can react with this 'intermediate' complex, due to the attraction between the δ^- charge on the water's oxygen atom and the δ^+ charge on the amino acid's carbon. This initiates a reaction which returns the enzyme to its original state, while releasing the other half of the original protein into the solution. Overall, the simple reaction shown at the top of figure 13.2 has taken place, but it has proceeded via a more complex multi-step reaction involving intermediate reactions between the protein and the surface of the enzyme. The enzyme provides an active chemical surface which encourages the reaction to proceed, essentially by providing an alternative route for the reaction.

You could, if you wished, consult biochemistry textbooks to discover many more chemical details of this reaction, and you could go on to learn about many more enzyme reactions and many more of the reactions which allow all sorts of proteins to perform their great variety of chemical tasks. At the end of your search, however, you would still be left with the same essential summary of protein activity which you already know: a protein folds up into a specific conformation which presents a particular chemical surface to the environment, allowing the protein to interact and react with chemicals in the environment in the ways that let it behave as an enzyme, a structural protein, a hormone, a transport protein, or whatever. That is the beautiful chemical simplicity behind the many powers of the proteins.

14   Leaves

The radiant energy of the sun travels out from that massive nuclear furnace in all directions, hitting many different objects as it goes. Some of it hits the tiny lumps of interplanetary and interstellar dust which are spread thinly throughout the near vacuum of space. Some of it strikes meteors and asteroids and comets, and some of it hits the airless surface of the moon, or is absorbed by the thick gassy atmosphere of Venus. All the planets of the solar system and all of their satellites receive their dose of solar radiation—radiation whose energy serves to push electrons of the chemicals it encounters up into excited states, up, in other words, into higher energy and rather unstable orbitals. The energized electrons soon fall back towards the atomic nuclei they surround, emitting energy as they go and thus assisting the chaotic dispersal of energy from the sun into the cooler, lower energy world around it.

In the green leaves of the plants of earth, however, something unusual and very significant happens. The solar energy first trapped by the raising of electrons into higher energy orbitals is not then recklessly dissipated away, but is instead channelled along the narrow chemical paths which allow the dead dust of the earth to be raised up into life. Leaves allow life to live.

We are all dependent on green leaves, and the chemistry of 'photosynthesis' which proceeds within them. The net chemical effect of photosynthesis is blissfully simple, and can be summarized by the following chemical equations:

$$\text{Carbon dioxide} + \text{water} \rightarrow \text{sugars} + \text{oxygen}$$

which can be represented symbolically as:

$$nCO_2 + nH_2O \rightarrow n(CH_2O) + nO_2$$

or, using glucose as a specific example of the type of sugar formed:

$$6CO_2 + 6H_2O \rightarrow C_6H_{12}O_6 + 6O_2$$
glucose

Plants use the carbon dioxide of the air and the water they draw up through their roots as the chemical raw materials needed to make sugars. These sugars are the basic carbon-containing raw materials of plant life. Their atoms, supplemented by other types of atoms gleaned from the soil, are then used for the construction of all the chemical structure of the plants. All animals are dependent on this chemical process within plants for two main reasons. First, all animals rely on plants as primary foodstuffs, because they live either by eating plants, or by eating other animals which are ultimately sustained by plants: all food chains depend on plants (or some bacteria which are also capable of photosynthesis) as the ultimate source of food. Secondly, the reactions of photosynthesis use up the carbon dioxide wastes breathed out by animals, and generate the oxygen gas which animals must breathe in.

Photosynthesis is essentially a reversal of the 'burning' of high energy foods which sustains animal life. When we eat sugar it becomes combined with oxygen to generate carbon dioxide and water, plus lots of useful energy as we considered in chapter 8. Plants reverse this process, regenerating the sugar we need by recombining water and carbon dioxide into sugars. Of course this transformation requires an energy input to make it proceed, since it is the reversal of a process which gives out a lot of energy. This is where the energy of the sun comes in, by providing the energy needed to push carbon dioxide and water up into the higher energy chemical configuration of sugar molecules and oxygen gas. The great chemical trick which proceeds in the leaf of a plant is the trapping of some of the energy of sunlight and its utilization to convert low energy raw materials into the higher energy products needed to sustain life. All life ultimately gets its energy from the sun, via the photosynthetic energy-trapping reactions which proceed in green plants (and the photosynthetic bacteria).

The way in which plants trap and use the energy of sunlight is very complex in its chemical details, but, as always, rather simple in principle. It depends in the first place on chemicals which can absorb some of the energy of the sun in a way that allows that energy to be trapped and then used, rather than wasted. The most important of these chemicals is known as chlorophyll, a rather complex molecule whose structure can be seen in figure 14.1. The long hydrocarbon tail serves to embed a chlorophyll molecule in the membrane of special membranous vesicles within plant cells known as 'chloroplasts'. The rather complex 'head' region is the part that absorbs some of the energy of sunlight.

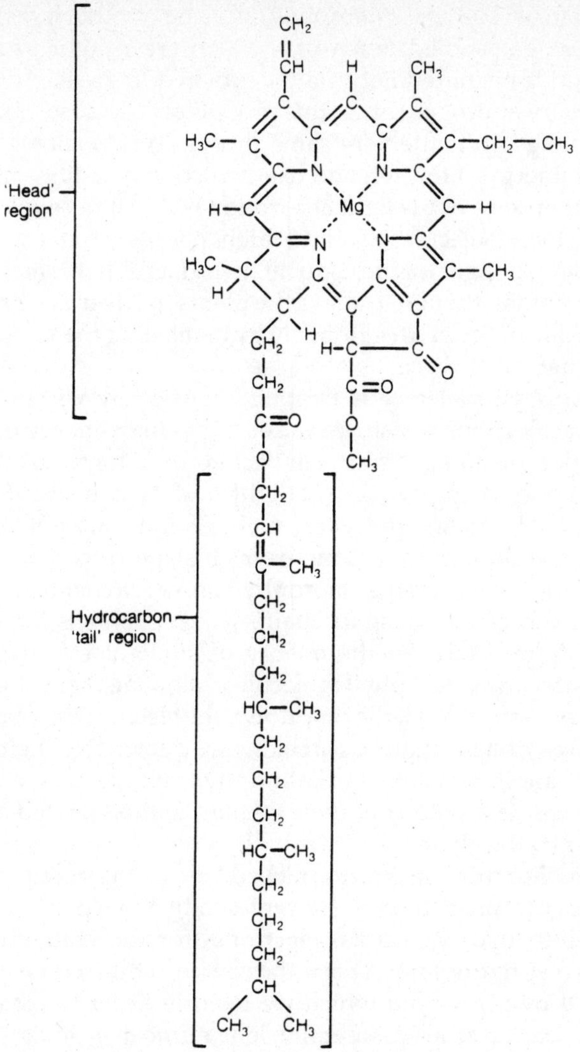

Figure 14. 1   The chemical structure of chlorophyll.

Chlorophyll is a green pigment—the pigment that is largely responsible for making plants green. The reason it is a pigment, the reason it is
coloured in other words, is because it absorbs part of the spectrum of
light arriving from the sun, absorbing light corresponding to colours
other than green more efficiently than it absorbs the green light. That is
why the light that is reflected from a leaf, or transmitted through it,

looks green to us: because chlorophyll (assisted by other less important pigments), has absorbed much of the non-green light to leave the reflected and transmitted light much enriched in green shades.

It seems very appropriate that this process of light absorption which makes plants green is the very same process which allows plants to capture the energy of the sun and use it to power the life of the earth. The real importance of the light absorption is that it raises electrons in the chlorophyll molecules up into high energy orbitals, and thus captures some sunlight for the use of life. The fact that this makes plants green is incidental, though it allows the plants to advertise their importance to the life of the earth gloriously by blanketing the earth in green wherever that life thrives.

If a chlorophyll molecule is floating about freely within a watery solution, the electrons which are raised up to high energy orbitals by the absorption of sunlight soon fall back to their lower energy levels with the energy being wasted as light and heat in the form of a 'fluorescence'. In plants, however, such electrons are not allowed to fall back to the orbitals they came from. Instead, proteins and other simpler molecules around the chlorophyll provide a chemical pathway, known as an electron transport chain, which channels the electrons away (see figure 14.2). So the energy of sunlight effectively ejects electrons out of the chlorophyll molecules, allowing them to pass from one chemical to another all the way down the electron transport chain. As you can see from the figure, part of the way down this chain they are passed on to another molecule of chlorophyll, allowing them to receive a further 'kick' from the energy of the sun and be passed along the second part of the chain.

As fast as electrons are being kicked out of the first chlorophyll molecules in this process, they are replaced by the 'splitting' of water molecules into hydrogen ions, electrons (for the chlorophyll) and oxygen gas (see figure 14.2). This is the source of the oxygen gas which is generated by plants and which we need in order to breathe. The hydrogen ions, or at least an equivalent number of hydrogen ions, eventually find their way into the sugars which you have been told are one of the end products of photosynthesis.

The electrons kicked out of the chlorophyll molecules eventually combine with hydrogen ions and a rather complex chemical known as $NADP^+$, to generate NADPH, in which the electron, hydrogen ion and $NADP^+$ are combined into one molecule; but during the passage of the electrons down the electron transport chain something very important happens. Hydrogen ions are pumped across the membrane in which the components of the electron transport chain are embedded. This happens in a rather complex way, but it is essentially an energy-

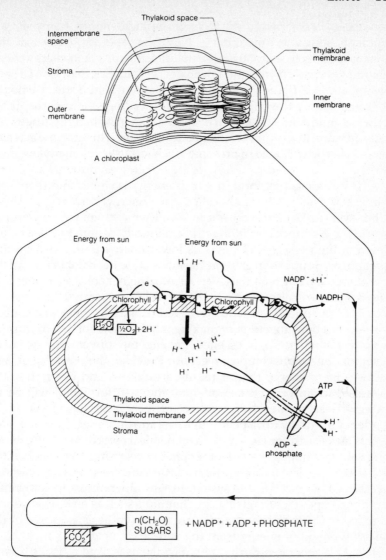

Figure 14.2 A summary of photosynthesis. The raw materials of the overall reaction—water and carbon dioxide—are identified by shaded boxes, while the products—sugars and oxygen—are in unshaded boxes. See text for details.

requiring process which is powered by the very transfer of the electrons along the chain. As electrons move along the chain they lose energy (apart from at the second chlorophyll molecule, of course) and some of this energy is trapped in the form of the high energy 'hydrogen ion gradient' across the membrane. Now, there is an enzyme in this membrane which can catalyse the manufacture of a high energy chemical known as ATP from its lower energy building blocks ADP and phosphate (see figure 14.2), but it can only do so if the reaction is accompanied by the effective transfer of hydrogen ions through the enzyme, in the direction shown, from a region of high hydrogen ion concentration to a region of lower concentration. This enzyme acts a bit like a water wheel, able to do some work (catalyse the manufacture of ATP) if it is powered by a flow of hydrogen ions through it.

ATP is the primary form in which energy is stored and then used within all cells. It is often called the 'energy currency of cells'. Whenever a reaction proceeds in a cell which requires an input of energy, it is (directly or indirectly) chemically coupled to the breakdown of high energy ATP into its lower energy products ADP and phosphate. Suppose an enzyme has to catalyse the combination of the low energy starting materials A and B into a higher energy product A–B. The enzyme will also be able to bind to ATP and catalyse its break-up into ADP and phosphate; and the structure of the enzyme will 'couple' the two reactions together by making it impossible for one to proceed without the other. So, in essence, the enzyme converts the two reactions into one more complex reaction. Provided the break up of ATP releases more energy than is required to join A and B together, the overall reaction of ATP degradation and A–B formation will be an energy-releasing one which can therefore proceed easily, driven by the tendency of energy to disperse out to the surroundings. Enzymes such as this make all the energy-requiring chemical reactions of living things proceed by coupling them to other energy-releasing ones, such as the break up of ATP, to make them part of energy-releasing reactions overall. Of course, for this system to work there must be a plentiful supply of expendable high energy chemicals such as ATP. That supply is generated by the clever chemistry of photosynthesis, powered by the relentless flood of energy from the sun.

The ATP generated during photosynthesis is used to make a great variety of energy-requiring reactions proceed, but the ones of primary importance are those which allow carbon, in its low energy form of carbon dioxide, to be incorporated into the high energy carbon containing chemicals of life. These reactions also consume the NADPH formed during photosynthesis, as indicated in figure 14.2. This figure probably looks rather complex to you, but it summarizes the chemistry of photosynthesis only in simplest broad outline. The workings of the chemistry of photosynthesis are far more intricate and complex than the figure suggests, and yet *overall* the chemistry is rather simple, the

atoms contained within water and carbon dioxide gas becoming rearranged into sugars and oxygen gas.

So we have accounted, in broad outline, for the conversion of carbon
dioxide gas and water into sugars and oxygen, and have seen how this
'reverse burning' process is powered by the energy of sunlight thanks
to some very complicated and clever chemistry—the chemistry of
electrons and energy which harnesses the sun as the engine of life on
earth.

15   Nerves

Your brain is the ultimate chemical reactor. A wet computer so accomplished that it is aware of its own existence and feels able to control its own behaviour. At least that is the aggressively materialistic view of your brain offered by modern science. Most chemists, and physicists and biologists, would be united in declaring that your head is full of chemistry, and chemistry alone. Of course many people think that there is much more to our brains, our consciousness and our thoughts than that. They talk of the brain as some sort of spiritual radio set, made from the hard stuff of matter but able to generate, sustain, transmit and receive mental phenomena of a still mysterious and mystical nature. It must be admitted that science is a long way from being able to explain how the chemistry of the brain can create a conscious mind and its memories and free will, if that is indeed what happens. What science can say is that the fundamental components of the brain appear to be nerve cells, many billions of which are linked up into a network which allows them to influence and be influenced by one another.

So what is a nerve cell, and what clever chemistry allows it to work? The question deserves a book, or several books, by way of an answer, but even in a short chapter like this I can reveal the bare essentials (see figure 15.1, and refer to it while reading the outline of nerve cell structure and function that follows).

Nerve cells, like all the cells of the body, are bordered by a thin membrane which has protein molecules and various other chemicals embedded in it. The centre of the membrane consists of stacked chains of linked atoms, each chain belonging to a type of molecule known as a lipid. The chains are known as hydrocarbon chains, because they consist of a row of carbon atoms which each has two hydrogen atoms bonded to it (apart from the final carbon, which has three hydrogens). These hydrocarbon chains give the cell membrane a vital property— they make it largely impermeable to the electrically charged particles we call ions, or, in other words, they make it an effective electrical insulator. The reason for this insulating property lies with the covalent bonds linking the carbon and hydrogen atoms. These bonds are not strongly polarized, since the nuclei of the carbon and hydrogen atoms attract the shared electrons to a rather similar extent (the atoms have similar electronegativities, in other words). So, throughout the region of stacked hydrocarbon chains, there are no significant regions of unbalanced full or partial positive or negative charge. This makes it very difficult for any charged particle to penetrate this region (i.e. such penetration takes a lot of energy), since the charged particle would need to move out of the watery environment beside the membrane, where its charge can interact with the partial charges on water molecules and the full charges on dissolved ions, and into the uncharged centre of the membrane where it would disrupt the Van der Waals bonds between the hydrocarbon chains. By acting as an insulator against the movement of ions across it, the membrane makes possible the creation of the electrochemical signals which we call nerve impulses. These signals depend on the controlled and assisted passage of ions through the membrane at appropriate times, to alter the distribution of charge across the membrane. The chemicals which give the assistance, allowing certain ions to break through the insulating barrier of the membrane, are proteins; and we must consider five main types of protein to gain an overall understanding of how nerves work.

One of these proteins is called the Na^+/K^+ pump. This is a protein which can bind to sodium (Na^+) ions inside a cell and transport them out, while at the same time binding to potassium (K^+) ions outside a cell and transporting them in, so it effectively pumps Na^+ out of cells at the same time as pumping K^+ ions in (see figure 15.1). Another protein, known as the K^+ leak channel. allows some of the K^+ ions to leak out of the cell much more quickly than Na^+ ions are allowed to leak in. The net effect of the activity of both these proteins is to make the layer of solution just outside a cell positively charged relative to the layer of cytosol (cell solution) just inside the cell; essentially because the K^+ leak channel lets K^+ ions leak out while no compensating Na^+ ions (or other positive ions) are allowed to leak in. All cell membranes possess this charge imbalance, but nerve cell membranes make special use of it.

A nerve impulse begins when a chemical known as a neurotransmitter (of which there are many different types) is released from one nerve cell to bind to 'receptor' proteins embedded in the membrane of another nerve cell. As a result of the binding of the neurotransmitter, the receptor protein undergoes a change in its structure which allows certain ions to pass through it. Many receptors allow sodium ions to flood into a cell in this state, the ions being pulled in by the electric attraction between themselves and the relatively negatively charged

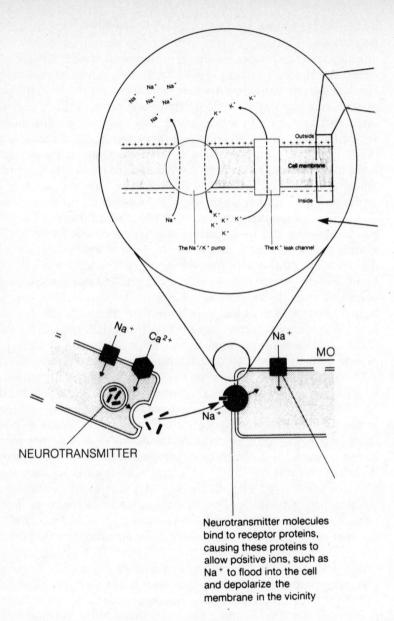

NEUROTRANSMITTER

Neurotransmitter molecules
bind to receptor proteins,
causing these proteins to
allow positive ions, such as
Na⁺ to flood into the cell
and depolarize the
membrane in the vicinity

Figure 15.1   Principles of the structure and function of nerve cells (see text for details).

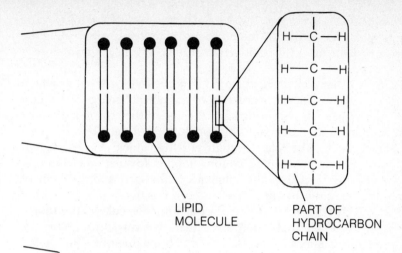

LIPID
MOLECULE

PART OF
HYDROCARBON
CHAIN

Cell membranes are electrically polarized, being positively charged on the outside and negatively charged on the inside, due to the action of the Na^+/K^+ pump and the K^+ leak channel.

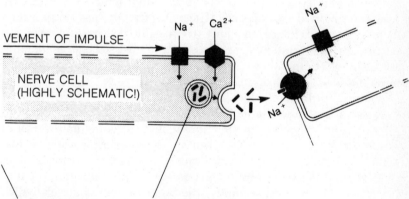

VEMENT OF IMPULSE

NERVE CELL
(HIGHLY SCHEMATIC!)

Na^+ channel proteins are induced by the membrane depolarization to allow Na^+ ions to flow through them. This reverses the polarization of the membrane nearby. The polarization reversal in one region induces neighbouring Na^+ channels to open, causing a wave of polarization reversal to spread along the cell membrane

When the nerve impulse reaches the terminal branches it induces membrane proteins to allow Ca^{2+} ions to enter the cell. The Ca^{2+} ions induce vesicles containing neurotransmitter molecules to release these neurotransmitters, allowing them to bind to the next nerve cell

interior of the cell. Of course, the entry of these ions tends to destroy the charge imbalance across the region of the membrane surrounding the receptor protein, since the positively charged Na^+ ions neutralize negative charge.

There are other proteins embedded in nerve cell membranes which are sensitive to any such destruction of the normal charge imbalance across the membrane. In response to the loss of the charge imbalance these proteins undergo a conformational change which briefly opens up a hole for Na^+ ions to pass through, and then quickly closes it again. While this protein is 'open', Na^+ ions flood in to such an extent that they reverse the normal charge imbalance of the surrounding region of membrane, making the interior of the cell positively charged rather than negatively charged relative to the outside. Molecules of this fourth important protein are studded at regular intervals throughout the nerve cell membrane, so the effect of one of them opening soon causes neighbouring ones to open, as they respond to the loss of the normal charge imbalance brought about by the opening of their neighbour. So the effect of a neurotransmitter binding to a receptor protein in a nerve cell is to initiate a pulse of electrochemical change which travels quickly along the length of the nerve cell. The main effect of this change is for a wave of reversed charge imbalance (technically 'polarization reversal') to spread across the membrane. This wave of reversed charge, rushing along a nerve cell membrane, is the electrical 'message' which we call a nerve impulse.

A nerve impulse really does spread as a discrete 'pulse' with regions of normal membrane charge imbalance both ahead of and behind it, because as one region of the membrane is undergoing its reversal of charge imbalance, preceeding regions are recovering their normal imbalance. This recovery occurs because, first, neurotransmitter molecules soon dissociate from the receptor proteins, returning them to their normal state; and secondly, the other protein channels which open up to allow Na^+ ions to flood in soon close up and then return to their normal state, all of which allows the Na^+/K^+ pump and the K^+ leak channel to restore normality (often assisted by various other proteins we need not consider).

That, in barest outline, is what a nerve impulse is and how it is created; but how does the passage of an impulse along one cell induce another cell, 'connected' to this first cell, to fire? When a nerve impulse arrives at the end of a nerve cell, it encounters yet another type of protein embedded in the membrane. This protein is induced, by the arrival of the impulse, to allow calcium ions (Ca^{2+}) to enter the cell. These calcium ions alter the chemistry of tiny membrane-bound vesicles found at the end of nerve cells, which contain lots of neurotransmitter molecules. This alteration induces the membranes of the vesicles to fuse with the cell membrane in a way which allows the neurotransmitter molecules to be released into the space between connected nerve cells which is known as the 'synapse' between the cells. The neurotransmitters diffuse across the synapse, bind to the receptor proteins embedded in the membrane of the nerve cell on the other side of the synapse, and so induce a nerve impulse to travel along the next nerve cell. Thus the transmission of an impulse along one nerve cell can induce the transmission of impulses along all the other cells to which a nerve cell is connected.

As you might expect, there are lots of complications and complexities. Most individual nerve cells receive signals, in other words they receive neurotransmitters, from many other nerve cells; and they also pass on signals, via neurotransmitters, to many other nerve cells. There are lots of different types of neurotransmitters, and while some of these encourage a nerve cell to 'fire' in the manner outlined above, others inhibit this firing. So in reality a nerve cell acts like a tiny chemical ballot machine. Its decision to fire or not to fire, or fire often or fire rarely, depends on the result of a chemical 'election' in which the votes are the many and perhaps conflicting signals it receives from other nerve cells. Other complications, complexities and subtleties are far too numerous and intricate for me to explore; but they all involve 'mere' chemical effects, similar to the most important ones outlined above, and each one explicable in terms of the interactions and reactions between chemicals as they are pushed and pulled by the electric force and the tendency of energy to disperse towards an even distribution.

So, even when we investigate the most complex things we know—our living brains—all we can find is chemistry. Chemistry that is built upon the laws of physics, of course, and which integrates to generate the phenomena of biology, but chemistry lies at the centre of it all, and is all we can find at the centre of it all.

Of course, that we can find only chemistry in the brain does not mean that other secrets do not lie awaiting discovery or perhaps forever hidden from our detection. As I have already said, science cannot yet provide a satisfactory explanation of the origin of a conscious mind, with all its thoughts and memories, from the observable chemistry that we find within the brain. All that has been found in the brain is an incredibly complex network of interconnected nerve cells, which pass nerve impulses along their lengths and pass a variety of chemical signals on to other cells which encourage or inhibit or otherwise modify the impulses which pass along these other cells. The general assumption of the scientific establishment is that consciousness, with all its mental sensations and thoughts and ideas, is simply what happens (somehow!) within a brain when its network of nerves undergoes specific and no doubt very complex patterns of firing. That assumption may be wrong.

When we 'remember' something our brain is presumed to generate a pattern of nervous activity which in some way resembles the pattern which was brought about by our initial experience of the thing we are remembering. Most theories of memory depend on molecular mechanisms that would allow the occurrence of a specific pattern of nervous activity to make it easier for the same pattern, or a similar one, to arise again. So something which has happened once in our brains would be more likely to happen again than something which has never happened; and we can somehow encourage the repeat performances of previous nervous activity to appear within our heads when we command them to.

Nothing, however, about the subtle details of our minds, our thoughts and our memories has been pinned down to the level of chemical equations and patterns of nervous activity which could be written in a book. We can write many equations and diagrams which describe in detail what happens when nerve cells fire and influence one another, but nothing yet which we could stand back and look at and say such things as: 'So that is the chemistry that makes you aware you are awake again; and these equations show how your brain creates the happy realization that it is a holiday; and this is how your brain evokes the memory of the smell of frying bacon and the anticipation of eating some with eggs and a cup of warm coffee.'

The mind is the secret inner chamber of biology which is as yet largely unexplored. Chemists, in one guise or another (as biochemists, neuropharmacologists and so on) are busily probing into all its secrets, since so far chemistry is all it has been shown to depend upon.

Is the materialistic view of the mind, as the direct creation of the chemical reactor we call the brain, a pathetically naive and presumptuous dismissal of a phenomenon whose secrets we can never hope to fathom? Is the mind some spiritual essence which resides in the fleshy brain thanks to the powers of the mystery known as 'God'? Or is there no God, no spiritual realm, no achemical wonders, but only a wondrous complexity of chemical reaction? We do not know. We may never know.

16   Electricity

Magic is easy with chemistry. A chemist can take some crystals of copper sulphate ($CuSO_4$), dissolve them in water to form a clear blue solution; and then add to this solution a powder of the bluey-grey metal known as zinc. After a little while the magic begins, for the mixture can be seen to contain brown deposits of copper metal, which grow larger as the zinc steadily disappears. The chemist could easily persuade a naive audience that they were witnessing the 'transmutation' of one element into another, the unrealized dream of the ancient alchemists (who would have been more interested in turning the zinc into gold, rather than copper). This transmutation is illusion, but the chemical reaction which mimics it is a representative of a very important class of reactions nonetheless. It represents the reactions which allow the energy stored within chemicals to be directly released as electricity.

What really happens when the zinc disappears and the copper appears is that the zinc atoms lose electrons to form zinc ions, while the existing copper ions gather up these electrons to form copper atoms. We can summarize this in equation form as follows:

$$Zn \rightarrow Zn^{2+} + 2e^-$$
$$Cu^{2+} + 2e^- \rightarrow Cu$$

where e^- represents an electron. The zinc ions pass into the solution, taking the place of the copper ions which are converted into solid deposits of copper metal. So all that has really happened is that *electrons have been transferred* from the zinc to the copper. The reason the electrons have been transferred is essentially that the combination of a zinc ion and a copper atom in water is a lower energy state than the combination of a copper ion and a zinc atom in water. So the dispersal of energy as the reaction proceeds ensures that it proceeds, overall, in the direction indicated.

All chemical reactions involve the rearrangement of electrons, but in many of them this rearrangement consists of the actual transfer of electrons from one chemical species to another. Reactions involving electron transfers are known as 'redox' reactions because they involve two coupled processes known as 'reduction' and 'oxidation'. A 'reduction' is a chemical process involving the gain of electrons, while an oxidation involves a loss of electrons. Obviously, all electron transfers must involve both a reduction and an oxidation since something must gain electrons during the transfer while something else loses them, so that is why such processes are known as redox reactions.

In the reaction we are considering the copper ions undergo reduction, or in other words they are 'reduced' to copper atoms when they gain two electrons; while the zinc atoms undergo oxidation, or are 'oxidized' to zinc ions when they lose two electrons.* The energy released during this redox reaction comes out as heat, dispersing away via the increased jostlings and vibrations of the chemicals all around. We can modify the reaction set-up, however, in a way that allows us to generate a flow of electrical current.

An electrical current through a metal wire is simply a flow of electrons as they jump from one metal atom to another. The electrons are pulled towards the positively charged end of the wire, and pushed away from the negatively charged end of the wire, by the electric force, and they are free to move from atom to atom for the reasons considered on page 28, when we discussed metallic bonding. So if we can somehow arrange for the electrons in our reaction to be transferred through a wire, rather than directly through the reaction mixture, then we will generate an electric current. This can be achieved very easily using the set up shown in figure 16.1.

The apparatus looks a bit complex, but in essence all that it does is separate the zinc atoms from direct contact with the copper ions which they must react with, although the two are connected by a metal wire and a copper plate through which electrons can pass. So zinc atoms can break up into zinc ions, which pass into the zinc sulphate solution, and electrons, which can be conducted away through the wire and into the copper plate. At the surface of the copper plate, the electrons meet up with copper ions which can combine with them to form a deposit of fresh copper on the existing copper plate. So the overall reaction between the zinc atoms and the copper ions can proceed as before. The

---

*The 'ox' in oxidation is derived from the word oxygen, but largely for historical reasons. Oxygen is one of the most effective elements at pulling electrons off other elements, and was one of the first such elements to be investigated in detail.

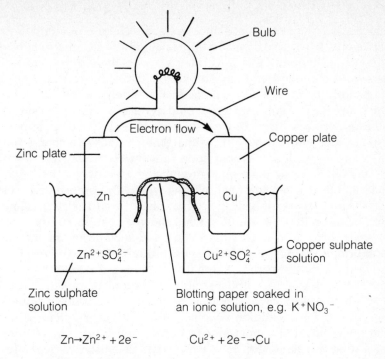

Figure 16.1 An electric cell (see text for details).

zinc plate will slowly disintegrate as it loses ions to the solution and electrons to the copper ions, while the copper plate will slowly grow larger due to the constant addition of fresh copper. Now, however, we are harnessing some of the energy released by the reaction in the form of an electric current which can be used, for example, to light a bulb. The total energy loss from the reactants remains the same, but some of it comes out initially in the form of light, which will often be of more use to us than heat. The reaction between zinc atoms and copper ions has been used to create an electric 'cell' or, in common usage, a battery (although strictly speaking a 'battery' should consist of a number of electric cells working together).

The zinc atoms and the copper ions are the only parts of the cell which actually react, albeit at a distance, to generate the electric current, but the other parts have their role to play as well. The wire and the copper plate provide a path allowing the electrons to flow between the reactants. The zinc sulphate solution provides a suitable 'sink' for the zinc ions, while the copper sulphate provides a 'reservoir' of copper ions. The piece of blotting paper soaked in a suitable ionic solution also has a vital role to play, by providing a path through which ions can slowly migrate to prevent any imbalance in electric charge from building up between the two solutions. To appreciate the need for this consider what would happen if this 'ion bridge' were not there. An excess of zinc ions would build up in the zinc sulphate solution, to which zinc ions are contstantly being added, making that solution positively charged; while an excess of sulphate ions would develop in the copper sulphate solution, from which copper ions are constantly being removed, making that solution negatively charged. The positive charge in the zinc sulphate solution would soon prevent electrons from flowing away from the zinc plate and through the wire, while the negative charge in the copper sulphate solution would also prevent these electrons from flowing towards the copper plate. So if the electrical imbalance between the two solutions were allowed to build up, the electric current would soon stop. This can be avoided by providing the moist ionic bridge linking the two solutions, through which positive ions can move towards the copper sulphate solution and negative ions can move towards the zinc sulphate solution. The ions are pulled across the bridge by the electric force between them, in just sufficient numbers to compensate for the addition of zinc ions to the zinc sulphate solution and the removal of copper ions from the copper sulphate solution.

So now you have seen how we can use chemistry to generate an electric current, not indirectly by releasing the heat energy which boils the water to make the steam which drives the turbines of a power station generator, but directly by separating the two halves of a redox reaction and allowing a wire to carry the electrons which must be transferred between them. As you might imagine, a great many different redox reactions are possible. Many are far more complex in their details than the simple example we have considered, but they are all equally simple in principle. The simple principle is that redox reactions involve the transfer of electrons from one chemical (or chemicals) to another (or others); and if we arrange for these electrons to flow through a wire we can generate a useful electric current.

The battery of figure 16.1 would certainly work, but it is not a very robust or practical device. Figure 16.2 shows you the essential chemical structure of one type of commerical battery. In this battery, lead atoms take the place of the zinc atoms of figure 16.1, while silver ions take the place of the copper ions. The ionic bridge consists of crystalline lead chloride, creating a 'solid' battery with no solutions able to slosh about or spill and leak out. There will be many water molecules present, however. Many batteries which are referred to as 'solid' or 'dry' batteries are in reality more likely to consist of moist pastes rather than perfectly dry solids. The water content varies, but

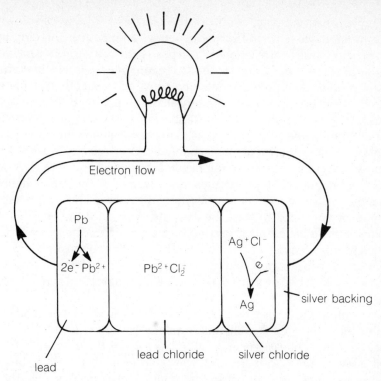

Figure 16. 2 *The chemical structure of one type of commercial battery (see text for details).*

they certainly do not contain any liquid solutions like the ones in figure 16.1.

The chemical reaction which powers the battery of figure 16. 2 is the transfer of electrons from lead atoms to silver ions, or in other words the oxidation of lead accompanied by the reduction of silver ions. The balanced chemical equation of the reaction is:

$$Pb + 2Ag^+ \rightarrow Pb^{2+} + 2Ag$$

Here we have a 'transmutation' that really would have interested the alchemists—the apparent conversion of lead into silver!

There are many different types of batteries available, all powered by different redox reactions, but even the most casual acquaintance with them reveals two main types—disposable batteries and rechargeable ones. In principle all batteries are rechargeable. When we recharge a battery we simply pump an electric current *into* it, rather than extract a current out of it. If we pump electricity into the battery of figure 16.2, for example, ensuring that we make the electrons flow in the *reverse* direction relative to their direction when we are using the battery, we can drive the chemical reactions which power the battery in reverse. We could force electrons onto lead ions to regenerate lead, and pull electrons off silver atoms to regenerate silver ions. We would then be able to extract useful energy from the battery all over again. In practice, many batteries, including that in figure 16.2, cannot be recharged very efficiently, simply due to the vagaries of their chemistry and their construction, but many batteries can be recharged very effectively indeed. The battery that provides the energy needed to start your car, by turning over the engine and supplying a spark to ignite the petrol, is a good example. You use its chemical reactions to release electrical energy whenever you need to start the car; and then once the engine has started some of its power is used to turn a generator (the 'alternator') which pumps electricity back into the battery, in the opposite direction from that in which it came out, to recharge the battery by reversing the reactions which had been used to turn over the engine and generate the spark.

Car batteries provide a nice example of the bidirectional possibilities of chemical change. The chemistry in the battery will move in the direction which releases energy, whenever that is the direction which best allows energy to disperse towards an even distribution; but if energy is pumped into the battery, as a charging electrical current, then the dispersal of energy *into* the battery will drive the chemical reactions in the other direction. In both cases energy is dispersing towards a more even distribution, but in one case that takes it out of the battery while in the other it takes it in.

I could write a whole book about chemistry and electricity, and all the various ways in which chemistry can generate electricity and in which electricity can affect chemical reactions; just as I could write a whole book about the chemistry of nerves, or leaves, or proteins, or genes, or carbon, or water, or air, or fire. I am restricting myself to the fundamental principles and powers of chemistry in this book, however, and this short chapter has already revealed the basic facts that underlie all examples of the generation of electricity from chemical reaction, or the powering of chemical reaction by electricity. Reactions in which electrons are transferred between chemicals can be harnessed to produce an electrical current by causing the electrons to be transferred indirectly, through a wire, rather than jumping directly between the chemicals concerned; and such reactions can be driven in reverse if we supply the electrical power needed to do so.

17 Soap

In chemistry, as in all science, each new discovery leads the way towards further discoveries. Learning how to initiate and control the chemistry of fire led our forebears towards the discovery of the chemistry of cooking; and when these two primitive chemical skills were combined they allowed the accidental discovery of chemicals that can clean. When animal fat drips from a roasting carcass onto the hot ashes of a wood fire the fat is able to mix with potassium carbonate in the ashes. If water is then poured onto the hot ashes to douse the fire, a chemical reaction can occur in which the fat molecules are hydrolysed (essentially, broken down by reaction with water) to release fragments which can then react with potassium carbonate to generate a rather special type of ionic substance, which we call soap (see figure 17.1). All soaps consist of chemicals in which a long hydrocarbon 'tail' region ends in a short electrically charged ionic 'head'. Someone, somewhere, a very long time ago, first discovered that the messy residue of the fire used to cook a meal could be mixed with water and used to clean up after the meal. That person discovered the chemical powers that we still rely on to wash ourselves, our clothes, and the dishes and pots and pans left over after our own meals. The chemistry that gave us cooking also gave us the ability to wash up afterwards! Washing with soaps and detergents is one chemical process which we all perform with great regularity. It is worthwhile briefly exploring the simple chemical principles upon which all our cleaning efforts rely.

Many modern soaps have a sodium ion in place of the potassium ion of the soap in figure 17.1, and some have hydrocarbon chains of different lengths; but they all share the same general structure of a long hydrocarbon tail region terminated by a short ionic head group. How does this structure allow the soap to clean? Most of the dirt we wish to clean away consists of greases, oils and small solid particles, or a mixture of solid particles embedded in greases and oils. The crucial feature of the chemical structure of all soaps is that their uncharged and

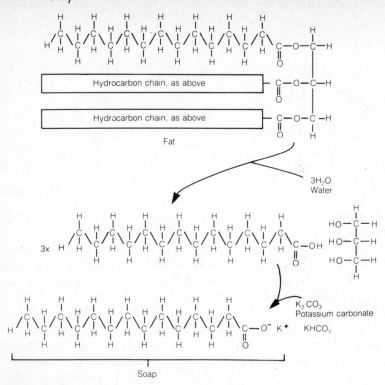

Figure 17.1   The formation of soap from fat.

unpolarized tail regions are hydrophobic (see page 140) and are therefore chemically attracted towards oil and grease and pushed out of
water. The electrically charged head regions, on the other hand, are
hydrophilic (see page 140), having a great chemical affinity for water.
So the chemical structure of soap gives it a split chemical personality,
which allows the soap to clean up greasy messes, as shown in figure
17.2.

When the soap is dissolved in water, it becomes a solution of the large
'soap' molecules with their charged heads, and whatever counter-ion
these heads were associated with. (Strictly speaking, the soap molecules are not true molecules, but are large molecular ions since they
carry a negative electrical charge on their head regions). The hydrophobic tails of the soap molecules will become buried in any greases
and oils wherever possible. For effective washing, the mixture of soap
and water and whatever is being washed must be agitated vigorously.
This lifts the greases and oils into the solution in the form of tiny
droplets which will immediately be stabilized by soap molecules

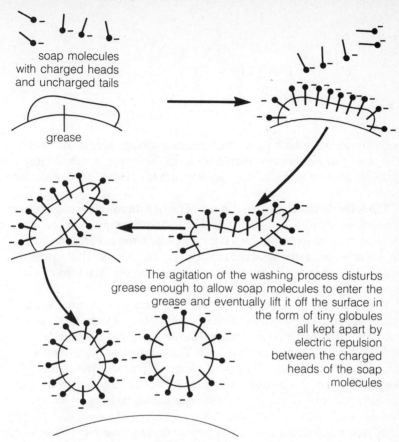

soap molecules
with charged heads
and uncharged tails

grease

The agitation of the washing process disturbs
grease enough to allow soap molecules to enter the
grease and eventually lift it off the surface in
the form of tiny globules
all kept apart by
electric repulsion
between the charged
heads of the soap
molecules

Figure 17.2   The cleansing action of soap (see text for details).

burying their tails in the grease and leaving their heads exposed to the
water. These droplets are prevented from coming together by the
electrical repulsion between the negatively charged heads of the soap
molecules, allowing the mixture of soap and grease and oil and
embedded dirt to be rinsed away and leave whatever was dirty now
fresh and clean.

Water on its own is not a very effecting cleaner, because greases and
oils do not dissolve in it. Soap effectively makes greases and oils behave
as if they were soluble in water, by allowing tiny droplets of them to mix
with water and so be lifted off any dirty surface and rinsed away.

Soap was a great discovery, and it became ever more pure and
refined as industrial chemists learned how to make the most effective
soaps, free of unwanted chemicals and by-products, and blend them

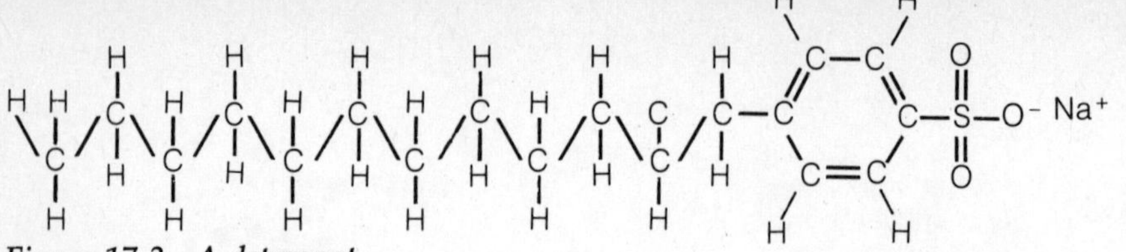

Figure 17.3   A detergent.

with chemicals which gave their soaps pleasant scents and colours. There was one major problem with soap, however, a subtle chemical problem known to the layperson through the terms 'hard water' and 'scum'.

The water available in many parts of the world contains large concentrations of dissolved calcium and magnesium ions, leached out of calcium- and magnesium-bearing rocks which the water flows through and over before arriving at the tap. The problem is that these positively charged ions combine with the large negative molecular ions of soap to form an insoluble precipitate known as scum.

The problem of scum led to the development of new synthetic cleaning substances which work in essentially the same way as soap, but which do not form the same insoluble precipitates when they meet up with calcium or magnesium ions. These new chemical cleansers are known as detergents. There are many different types, although a typical modern one is shown in figure 17. 3. Just like soap, a detergent consists of a long hydrophobic tail region, able to bury itself in grease and oil, and a short hydrophilic head which tends to remain in contact with water. Detergents are our synthetic improved versions of the natural soaps which were first discovered in the warm ashes of the fires of our most distant human ancestors.

18 Materials

We apparently share common ancestry with all the other living things of the earth—we even look rather similar to some of them—yet an alien observer would instantly recognize one dominant characteristic distinguishing us from all other life: our mastery of the earth's materials.

Human hands have dug up the many materials of the world, human minds have pondered on their properties, and the hands and minds have fashioned the raw materials into clothes and tools and machines and much more besides. The first tools and machines led the way to the development of new tools and machines with which to dig up and process and manipulate materials in ever more complex and useful ways. The materials were purified, studied and modified in a global effort which created the vast field of modern technology known as materials science, which is, of course, merely a branch of chemistry.

The first materials humans gathered and examined and exploited were natural substances readily available in the world around them. They used animal skins and furs as clothes; branches, leaves, mud and stone as building materials, and so on.

Soon, they discovered that the chemistry of fire could be used to release many of the versatile materials we call metals from their ores. One common ore of iron, for example, is a form of iron oxide known as 'hematite' whose formula is Fe_2O_3. The precious iron of hematite is locked up in chemical combination with oxygen, since free iron naturally reacts with oxygen to generate various forms of iron oxide (including the rust into which the iron of our motor cars slowly but inevitably turns). Iron can be released from hematite, however, if the ore is brought into contact with hot carbon, such as the charcoal left behind after a fire. The carbon reacts with the iron oxide to generate carbon monoxide and carbon dioxide, effectively stealing the oxygen from the iron of the ore to leave the free metal available for our use. As soon as the metal is liberated from its ore it begins to react with the oxygen in the air and turn slowly to rust, but if we treat it carefully it can

169

remain predominantly as free iron long enough for us to put it to good use.

We have learned how to release a wide range of metals from their ores, and then how to mix different metals to form alloys with further distinctive and useful properties. The chemical knowledge of the earth's materials gained over countless generations placed the last few generations in a position to begin manufacturing unnatural synthetic materials, especially all of the man-made fibres and plastics so evident in the modern world. In this chapter we will briefly examine a small selection of natural and synthetic materials to gain a good overall insight into the chemistry of some of the most important materials around us.

Rock

Rocks are the naturally occurring hard solids of the earth, and they include many different chemical compounds. They can be divided into three main categories: igneous rocks, sedimentary rocks and metamorphic rocks.

Igneous rocks are those which have cooled and solidified from a hot, molten, i.e. liquid, state. They are most commonly formed from cooling volcanic lava, parts of which can turn into solid rocks either before or after the lava becomes exposed on the earth's surface. Like most rocks, they tend to consist either of ionically bonded crystals or of giant networks of polar covalently bonded atoms which can behave like crystals or like glasses. Remember, however, that the distinction between ionic and polar covalent bonding is a rather artificial one, with most bonds possessing some characteristics of both these ideal extremes. This is particularly true of the bonding in many rocks, so most rocks are best regarded as giant networks of atoms and/or ions all held together by polar covalent and/or ionic bonds, with the true character of the bonds often rather mixed and unclear. In all cases, of course, it is the force of electric attraction between electrons and protons that holds the rocks together.

What atoms and compounds do igneous rocks actually contain? The answer is a bewildering diversity, but ten compounds very commonly found in such rocks are SiO_2, TiO_2, Al_2O_3, Fe_2O_3, FeO, MnO, MgO, CaO, Na_2O and K_2O. This list gives you a good impression of the chemical variety of rocks, but notice that all of these compounds contain oxygen atoms. Oxygen is believed to be the second most abundant element in the earth, by weight (iron is the most abundant), with much of it being bound up into solid rocks. The free oxygen of the air is actually a bit of a chemical peculiarity, created by the oxygen-releasing reactions of photosynthesis. Oxygen atoms are highly electronegative, and so they tend to react with a wide range of other elements in a way which gives the oxygen atoms a share of these other elements' electrons. The strong electronegativity of oxygen is why it reacts so vigorously with many chemicals in the range of reactions known as 'fires', and why so much of it is contained within the hard rocks of the earth.

Most igneous rocks consist of a complex mixture of compounds such as the ones listed above. The well known igneous rock granite, for example, contains 70 per cent SiO_2, 14 per cent Al_2O_3, 4 per cent K_2O, 3.5 per cent Na_2O and smaller amounts of many other compounds.

Sedimentary rocks are formed from compounds that have been carried in water and have then precipitated out of the water to form sediments. These sediments tend to build up very slowly, layer by layer, over the years, with the layers eventually being compressed together into solid rock by the weight of the new sediments accumulating above them. So these are rocks that form at the bottom of the seas and lakes and rivers. They are made from chemicals that have been released into water by the weathering of existing rocks on the land. So the water that continually washes and wears at the rocks of the earth carries away the raw materials for the construction of new sedimentary rocks.

The chemical composition of sedimentary rocks also varies widely, although their most common constituents are the compounds SiO_2, Al_2O_3, CaO, FeO, Fe_2O_3, CO_2, $CaCO_3$, MgO, K_2O and Na_2O. So they contain compounds also found in igneous rocks, as you might expect since they are often formed by the weathering of igneous rocks. Sandstones, limestones and shales are all common examples of sedimentary rocks.

Metamorphic rocks are those whose chemical structure has been modified (metamorphosed) by exposure to high temperatures and pressures in the interior of the earth. The common sedimentary rock limestone, for example, can be metamorphosed into marble by processes which remove water, carbon dioxide, sulphur and silicon dioxide from the limestone. A great variety of different metamorphic rocks exists, just as there is a great variety of igneous and sedimentary rocks, but all metamorphic rocks are formed from the chemical constituents of other rocks, altered in various ways.

I cannot be too specific in a book such as this, since to be specific would require many more words than I wish to use. I hope I have given you some inkling, however, of the chemical form and variety of the hard rocks of the earth. These are the solids which formed the first truly durable materials which could be manipulated and exploited by mankind; and of course we continue to make great use of their strength and durability in the modern world.

Iron

The earth contains more iron atoms than any other type. They form most of the solid core and hot molten outer core at the centre of the earth; they are found in a variety of 'iron ores' in the rocks of the earth, chemically bound into compounds such as Fe_2O_3 (hematite), Fe_3O_4 (magnetite), $FeCO_3$ (siderite) and FeS_2 (iron pyrites); they serve vital roles in living organisms, such as providing the iron ions to which oxygen molecules become attached in order to be transported from your lungs to every cell of your body; and they are released from their ores by industry and used as structural materials in the form of pure iron, a wide variety of steels, and various other alloys as well. So we depend on iron atoms and ions in many ways, as components of the earth, its life, and human industry and technology.

Iron atoms each contain 26 protons, 26 electrons, and usually 30 neutrons. For reasons related to their electron arrangement the atoms can readily form two types of iron ions—Fe^{2+} and Fe^{3+} depending on the chemicals they react with and the conditions in which they react, and they can also participate in polar covalent bonds and the metallic bonding that holds together iron metal. This all gives iron a great chemical versatility, allowing it to be part of many different chemical compounds.

As a material, iron is largely used by us in the form of relatively pure iron, or as various alloys of iron, with an alloy defined as a homogeneous mixture of iron and one or more other types of atom. In order to make use of the strength and versatility of iron, it must first be released from its naturally occurring ores. This is nowadays done in blast furnaces, in which a mixture of iron ore (usually Fe_2O_3), coke (essentially carbon—C) and limestone ($CaCO_3$) is poured into the top of the furnace to meet a blast of hot air (containing oxygen—O_2) rising upwards. The mixture is ignited, causing the coke to burn to generate carbon monoxide (CO) and an intense heat. The iron ore is converted into free iron by a mixture of the two reactions shown below:

$$Fe_2O_3 + 3CO \rightarrow 2Fe + 3CO_2$$
$$Fe_2O_3 + 3C \rightarrow 2Fe + 3CO$$

The temperature in the heart of the furnace approaches 2,000 degrees centigrade, well above the melting point of iron, which is 1,535 degrees centigrade. So the iron is formed in a molten state, allowing it to gather in a pool at the bottom of the furnace which can be tapped off and allowed to cool into solid 'cast iron' (also called 'pig iron'). The cast iron coming out of a blast furnace is rather impure, and so further chemical processing is required to remove the impurities. These processes do not remove all of the carbon present, however, and so eventually they yield a form of iron which contains a little over 1 per cent carbon atoms. This material is known as carbon 'steel', since steel is the name given to alloys containing a mixture of iron and between 0.1 and 1.5 per cent carbon atoms. It is possible to treat cast iron in a way that removes almost all of the carbon, to generate a virtually pure form of iron known as wrought iron; but the carbon atoms of steel actually make it a more versatile and useful material than pure iron, so most iron released from iron ore eventually ends up in the form of steel.

Steel is our major industrial metallic material, and there are many different types of steel containing differing amounts of carbon and other atoms and all differring in properties such as toughness, malleability and resistance to corrosion. By 'fine-tuning' the basic material strength of iron through the addition of small amounts of other atoms, humanity has produced a range of steels to suit almost any structural task. The first humans to notice the release of iron from its ores in the ashes of their fires started quite a trend—nowadays about 700 million tons of glowing molten iron comes rushing from the fires of industry every year.

Rubber

A wounded rubber tree slowly oozes forth a tacky liquid which, after suitable processing, yields rubber. It is fascinating to speculate on the discovery of the useful properties of nature's own materials. Who first wounded a rubber tree and discovered the usefulness of the goo that emerged? Of course the discovery was probably made independently many times and in many places, just like the discovery of fire, and soap and iron; but someone must have been first. Whoever was first to investigate the liquid seeping from a wounded rubber tree was the founder of a vast modern industry. If they had been able to patent their discovery, as modern biotechnologists can theirs, their descendants would now be very rich!

The main characteristic of rubber which makes it useful is its elasticity. A piece of rubber can be stretched or compressed, only to spring back to its original shape as soon as it is released. So it is a bouncy, flexible, elastic substance, which also has the very useful property of being waterproof, and all these characteristics are due to the chemistry that holds rubber together.

Chemically, rubber is rather simple. A piece of rubber is composed of many millions of large rubber molecules which each consists of thousands of carbon and hydrogen atoms linked into long chains, as shown in figure 18.1. This figure reveals that rubber is a polymer made by the linkage of many identical small molecules known as isoprene molecules. These are combined within the rubber tree (and to lesser extents within many other plants as well) to form long, kinked and coiled molecules of rubber which are, on average, 5,000 isoprene units long. The kinking and coiling of the rubber molecules is the key to rubber's elasticity. It means that any piece of rubber consists of large numbers of molecules all entangled together. A stretching force tends to straighten out the entangled chains a little, but imposes the strains on the molecules that tend to pull them back towards their original arrangement as soon as the stretching force is released. A compressing force tends to kink and entangle the molecules a little more, but creates the strains that will push them back towards their original state when the force is released. So rubber's ability to be stretched and compressed depends, ultimately, on the ability of its chemical bonds to be flexed away from their natural angles, allowing the molecules either to straighten or to kink a bit more than they do naturally. In either the stretched or compressed state, however, electric forces of attraction and replusion between the rubber's electrons and nuclei create the forces needed to pull the rubber back into shape. It is waterproof because it consists of a dense tangle of hydrocarbon chains which, for reasons already considered, have no chemical affinity for water. The structural strength of the rubber, its ability to hold together rather than snap or crumble when stretched, for example, depends on the entangling of the long rubber chains, preventing them from being completely pulled apart.

Actually, natural rubber is not particularly strong, but in 1839 the American Charles Goodyear found a rather simple way to make it much stronger. He found that heating natural rubber with sulphur caused individual rubber molecules to become linked together by two sulphur atoms joined to themselves and to two rubber molecules by three covalent bonds (see figure 18.1). This created what is known as vulcanized rubber, which is much stronger, since the rubber molecules are now bonded to one another, rather than being merely entangled together. Goodyear's name still decorates many of the motor vehicle tyres of today, in testimony to the importance of this discovery.

So, once again, we find chemists discovering useful properties of the natural materials of the world, and then improving on them by

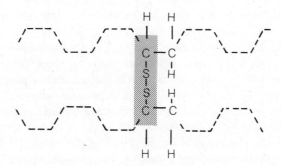

Isoprene

A rubber molecule

A sulphur 'bridge' in vulcanized rubber

Figure 18.1 Rubber.

incorporating subtle chemical modifications. Eventually, the process of exploiting, studying and modifying natural materials tends to give chemists the ability to mimic nature by making synthetic materials which are often modelled on natural ones, but with properties specifically tailored to suit human needs. This has happened with rubber, since most modern 'rubbers' are synthetic rubbers made by joining together monomer units different from natural isoprene. They are known as synthetic rubbers because they share some chemical similarities with natural rubbers and they are, like natural rubber, waterproof and elastic; but they also have much improved properties of toughness, durability, flexibility, elasticity and so on, compared to natural rubber.

The best examples of materials chemists' abilities to learn from nature, and then improve on nature, come from a slightly different field of materials science: the field of man-made fibres.

Man-made fibres

Nature makes great use of fibres, especially in the living world. Plants and animals contain a diversity of fibrous materials, largely composed of chains of natural polymers wound around one another. As humans began to learn how to exploit the materials of the living world in more subtle ways than simply wrapping themselves up in skins and furs, they discovered natural fibrous materials that could be spun into thread and woven into cloth. Wool, cotton and silk are the most notable examples, and clothes made of these materials have kept people warm for centuries. Nowadays, however, we are more likely to be warmed and decorated by clothes of man-made fibre or of some mixture of man-made fibres and wool, cotton or silk.

Man-made fibres are organic polymers, or in other words are composed of long chain-like molecules made by the linkage of many smaller 'monomer' molecules in which carbon atoms play a central role. Just like natural polymers, most man-made fibres depend on chains of linked carbon atoms, sometimes interrupted by other types of atoms, for their structural integrity.

Many different man-made fibres are available, each with its own characteristic properties. The names, or at least the common names, of some of them should be familiar to you: nylon, polyester and acrylic, for example.

Different types of each of these general classes of man made fibres exist, but typical ones are shown in figure 18.2. Nylon can be made when molecules of diaminohexane and adipic acid undergo condensa-

a Formation of nylon

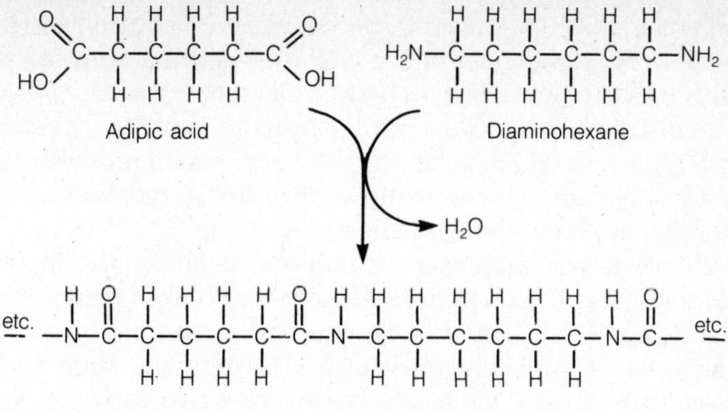

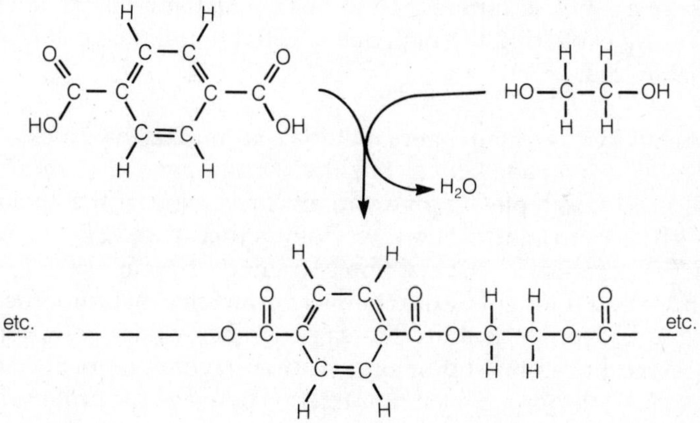

b Formation of polyester

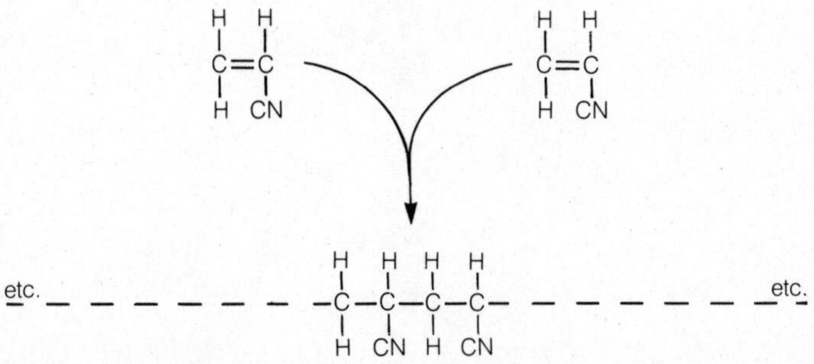

c Formation of acrylic

Figure 18.2 Man-made fibres.

tion polymerization to create a polymer chain in which the two monomer units alternate (see figure 18.2(a)). Polyesters are also formed by condensation polymerization between monomers such as those in figure 18.2(b). Acrylic fibres are formed by a different type of polymerization reaction, in which small monomers containing double bonds merely add on to one another as shown in figure 18.2(c), with no water or other chemical being released as a by-product.

The chemical raw materials for man-made fibres, the monomer molecules in other words, must still be gathered from nature, or else prepared from other chemicals gathered from nature, but the fibres themselves are completely unnatural. The ultimate source of the monomer units is usually oil, which contains a rich variety of simple organic chemicals in addition to the ones we use to propel and lubricate our vehicles. Oil, of course, is formed from the remains of long-dead plants and animals, so as we wear our polyester shirts and acrylic jerseys we are not so different from our ancient ancestors—warming ourselves by exploiting the materials created within ancient animals, just as they did.

There are, of course, many more materials around us than rocks, iron, rubber and man made fibres; but these have been chosen as four representative examples to give you an impression of the chemistry that holds the substances of our world together. I could have chosen plastics, or ceramics, or natural fibres, or modern composites . . . The list of materials is long, but the chemical principles that underlie their structures and properties are the same. Materials are composed of molecules and/or atoms and/or ions, all held together, ultimately, by an 'electric glue': the force of electric attraction between the protons in the nuclei of atoms and the electrons in their orbitals around them.

19   Medicines

We are living in a golden age of medicine. The art and science of medicine has, of course, been practised for centuries, but only in this twentieth century has it begun to win major triumphs over disease and disability. Vaccines are available to prevent a wide range of potentially lethal infections; many of us would not be alive today without the bacteria-fighting action of antibiotics; nobody need suffer the full agonies of persistent pain any more, for we have drugs that can abolish or at least drastically reduce it; and a host of other drugs are available to cure and relieve all manner of ailments. Many modern drugs may be overused or used unnecessarily, many can be harmful as well as helpful, but whatever their drawbacks and dangers the drugs of the modern world are one of the major reasons why the world is a safer and more pleasant place to live in than it used to be.

By now, you should not be expecting a comprehensive review of the chemistry of drugs in this chapter—that would require many chapters, or perhaps many books. In fact, you are not even going to be offered a quick summary of the chemistry of drugs. Instead, I will focus on one new class of drugs currently being developed by all the major drug companies, which I have chosen to serve as an illustrative example of the chemistry of medicines in general. You learned in chapter 13 that the activities of living bodies are controlled by the chemicals we call proteins, including the enzymes that catalyse the chemical reactions of life. When we fall ill, the illness often involves problems with our proteins; and in fact the vast majority of drugs either act by binding to and affecting specific proteins, or else they are proteins themselves. The central importance of our proteins in health and disease will be found at the heart of the drugs I have chosen to consider here—a range of chemicals which might yield some of the most important and widely used drugs of the future.

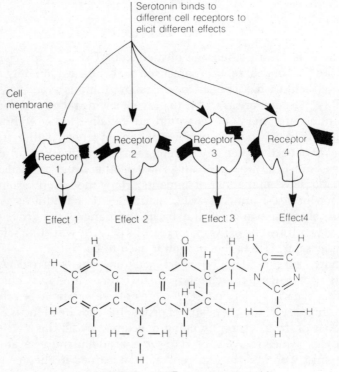

5-Hydroxytryptamine (5-HT) (Serotonin)

Serotonin binds to different cell receptors to elicit different effects

Cell membrane

Receptor 1   Receptor 2   Receptor 3   Receptor 4

Effect 1   Effect 2   Effect 3   Effect4

GR38032F - a 5-HT$_3$ receptor antagonist

Ketanserin - a 5-HT$_2$ receptor antagonist

Figure 19.1   Serotonin and serotonin-related drugs (see text for details).

Serotonin—a clarifying enigma

There is, within all of us, a single simple chemical which influences an astonishing number of the biochemical processes that keep us healthy. It is known as serotonin, although its proper chemical name is '5-hydroxytryptamine' (often abbreviated to 5-HT). As you can see from figure 19.1, serotonin is a small molecule of just 25 atoms, and yet this one chemical seems to play a crucial role in an incredible variety of processes within the body. It is believed to be involved (amongst other things) in controlling our moods and emotions and indeed our sanity itself, our sexual behaviour, our perception of pain and nausea, our food intake, the regulation of the temperature of the body and our patterns of sleep and wakefulness. Since it is crucial to such processes in health, disturbances in the processes due to disease can often involve disturbances in the normal activities of serotonin.

When serotonin was discovered, 40 years ago, the American Irvine Page (one of the discoverers) christened it 'the enigmatic hormone' since nobody really new how important it was or exactly what it did. Scientists are still confused by many of its actions, but now are convinced of its vital importance. One of the main enigmas was the question of how one simple chemical can play so many roles, and be involved in so many disease processes. The answer is now becoming apparent, and it lies in the role of serotonin as one of the body's 'transmitter' substances, which serves to allow cells to communicate with and influence one another in many different ways. To understand how this role gives serotonin its remarkable versatility you need to know a bit more about how biological transmitters work.

The body contains many chemicals which can be described as transmitters, all being released from specific cells in order to affect other cells. The way in which they bring about their effects is to become bound to large 'receptor' molecules belonging to the target cells, and usually found exposed on the surface of these cells. Most of these receptors are proteins, perhaps modified by the addition of other types of chemical, and each receptor contains a 'binding site' on its surface to which an appropriate transmitter can become bound. The binding site is merely a particular three-dimensional array of atoms and/or ions which are part of the receptor, and which form a cleft into which a particular transmitter can become bound in a highly specific manner. The binding of a transmitter to a receptor stimulates the receptor into modifying the chemistry of the cell it belongs to, by acting as an enzyme, for example, or in some other way. The vital points are that for each transmitter there is at least one receptor, often more, and when the transmitter binds to the receptor it stimulates the receptor into bringing about a very specific chemical change within its cell.

Some transmitter substances are the neurotransmitters which we met in chapter 15. These are released from nerve cells to bind to receptors on neighbouring nerve cells and stimulate or inhibit or modify the target cells' nervous activities. Other transmitters are called hormones, which are released from specialized cells in the body to travel around the body, usually in the bloodstream, and affect a variety of cells at sites which can be far from where the hormones are released. The third major class of transmitters are known as 'tissue factors', which act a bit like 'local' hormones whose effects are restricted to cells rather close to their sites of release; and other types of transmitters, in addition to these 'big three', are known. The distinction between the types of transmitters is rather artificial, invented by humans to try to make sense of the often bewildering chemistry of the body. Some chemicals can cross the category boundaries and act as more than one type of transmitter, and none more so than serotonin. It is certainly a neurotransmitter in the brain, and a hormone and tissue factor at various sites in the body; and the full range of its powers is not yet known. The secret of its versatility, however, clearly lies in the wide variety of cell receptors to which serotonin can bind; and remember, each type of receptor evokes a different response in its cell when serotonin becomes bound to it.

There are at least three main types of serotonin receptors, known as 5-HT_1, 5-HT_2 and 5-HT_3 receptors (remember 5-HT is the abbreviation for 5-hydroxytryptamine, the proper name for serotonin). Each of these main types is being found to consist of several subtypes (such as 5-HT_{1A}, 5-HT_{1B}, and so on) as the receptor variety becomes more fully characterized . So serotonin is a single simple messenger which can evoke many different responses in a cell, depending on which type of serotonin receptor is carried by the cell. There lies the answer to the many different roles this one chemical seems to play within the body.

As scientists seek to uncover the mysteries of serotonin's many activities, they find that one of the best ways to discover the function of each type of receptor is to investigate the effect of other chemicals which can bind to serotonin receptors and mimic or block all or some part of serotonin's normal effects. All such agents are potential drugs. They can be highly selective about the specific type of serotonin receptor they will bind to, and they fall into two main classes: chemicals which bind to serotonin receptors and mimic serotonin by stimulating either its natural response or some modified response are known as receptor 'agonists'; while those which bind to a receptor and block its normal response, perhaps by physically preventing serotonin from binding, though perhaps in some other way, are known as receptor 'antagonists'.

Some agonists and antagonists of our serotonin receptors are already on the market as drugs as I write. These include a drug known as 'ketanserin' (made by Janssen Pharmaceutica) which combats high blood pressure (see figure 19.1), and 'buspirone' (from Bristol–Myers) which lessens anxiety. Many more are in various stages of development ranging from early animal studies to late clinical trials, and many important new drugs are expected to emerge from such efforts over the next few years. Let's look at a few of the most hopeful areas of endeavour.

Various drug companies are testing antagonists of $5\text{-}HT_3$ receptors which can reduce the vomiting associated with anti-cancer chemotherapy. This distressing side-effect is a major factor causing patients to withdraw from such therapy, so if it could be prevented then cancers might be cured which would otherwise prove fatal. One of the compounds being tested is made by Glaxo, and currently known only as GR38032F (see figure 19.1). Dr David Cunningham of St Mary's Hospital Medical School in London has described his early clinical trials of this compound as 'quite incredible', enthusing about the 'spectacular success' of the drug when compared with those currently available.

Clinical trials are also proceeding with $5\text{-}HT_1$ agonists for the treatment and prevention of migraine, and early reports indicate encouraging success. It has been shown, for example, that $5\text{-}HT_1$ receptor agonists such as a Glaxo drug known as GR43175 can provide effective relief of migraine headache within 30 minutes and with no serious side-effects.

$5\text{-}HT_3$ receptor antagonists have recently been shown to be powerful anxiety-relieving agents, while drugs that modify the sensitivity of $5\text{-}HT_1$ and $5\text{-}HT_2$ receptors are showing promise in the relief of depression.

$5\text{-}HT_{1A}$ receptor agonists and $5\text{-}HT_3$ receptor antagonists may be capable of treating schizophrenia, since they can reverse some of the biochemical abnormalities in the brain which are most associated with the disease. Schizophrenia is associated with the overactivity of nerve cells, in specific parts of the brain, which use a chemical called dopamine as a neurotransmitter. In studies on laboratory animals, $5\text{-}HT_3$ receptor antagonists can prevent dopamine induced hyperactivity when injected into the relevant regions of the brain; while $5\text{-}HT_{1A}$ receptor agonists can actually prevent the release of dopamine from nerves in these regions. Many drugs are currently used to treat schizophrenia, though with only limited success, and most of them act as antagonists of dopamine receptors. Antagonists and agonists of serotonin receptors offer a new and perhaps ultimately more effective approach.

Serotonin-related drugs, generally antagonists or agonists of serotonin receptors, may also eventually be used to control our appetite, sleep patterns and various other body functions as well. Since serotonin is involved in so many important processes in the body, serotonin-related drugs may be able to correct many of the abnormalities in these processes which we know as diseases.

The great hope for the new class of serotonin-related drugs under development is that they will be much more specific agents than most current drugs: that they will be able, in other words, to interact very precisely with the specific class of receptors implicated in any particular disorder, while leaving most of the body's other receptors well alone. This should make these agents more powerful in lower doses, and also much less likely to induce harmful side-effects, compared with current alternatives. The first results from clinical trials suggest that these hopes are realistic. Dr Brian Jones of Glaxo's serotonin research group, for example, has said that the 5-HT$_3$ receptor antagonists he is working on promise to be 'far superior to the drugs presently used to treat mental disorders, by being effective therapeutic agents without side-effects.'

Effective therapeutic agents without side-effects are the research pharmacologists' ultimate dreams. Only time will tell us how close to that ideal serotonin-related drugs can come. Some of the hopes may turn out to have been unrealistic, but it does seem likely that by carefully dissecting out the fine details of serotonin's actions on the body doctors may be provided with many new drugs able to treat important illnesses with the chemical equivalent of tweezers rather than sledgehammers.

The contents of this short chapter will probably be out of date by the time the book appears, and will date further as time goes by and as some of the drugs mentioned either reach the market and succeed or perhaps fail to live up to their early promise. For our purposes that does not matter. The chapter should have given you a useful insight into one class of drugs, and the state of their development, regardless of how far that development process has progressed by the time you read it. It should have helped you to appreciate that all drugs are chemicals which interact in specific ways with the natural chemicals of the body. The search for cures to the many ills that afflict us is a search for a better mastery of chemistry.

20   Epilogue

I lie in bed, contemplating chemistry. I think of the publication called *Chemical Abstracts*, which contains a comprehensive list and concise abstracts of the scientific papers which record humanity's knowledge of chemistry. It is enormous, consisting of rows of hefty volumes arrayed on long library shelves which are stacked up from the floor to the ceiling of the chemistry library. A ladder must be climbed to reach the highest volumes, and the shelving spans tens of yards before any other books are found; and yet it all contains merely the briefest possible summary of the knowledge held in journals which would occupy many times the space. I think of the countless millions of words of information; and then my mind recalls the simple chemical process of distillation, in which the most precious essences of a complex mixture can be boiled off, to be cooled, collected and treasured.

In metaphor mode, my mind imagines the volumes of *Chemical Abstracts* tumbling from their shelves and into a vat for distillation. I see a confusing swirl of formulae and phrases and equations as the pages open and crumple and tear and turn liquid and begin to bubble. Then the essence begins to rise, astonishingly slight, astonishingly pure, to be cooled and collected as a single drop that spreads out across a clean page of paper; and in my mind I see the drop imprinting the page with these few words—the central principles of chemistry, and truly things to treasure:

Chemicals are made of atoms and molecules and ions of matter; and these are made of protons, neutrons and electrons in differing arrays.

The electric force attempts to draw the protons and electrons towards one another, and to cause the protons to be repelled from protons, and the electrons to be repelled from one another also.

The efforts of this force can be resisted by the energy which flows through the microworld, always moving towards a more even distribution overall, and causing the atoms and molecules and ions to be constantly in motion and frequently in collision.

One result of these collisions is to make chemical reactions proceed, reactions in which existing chemical bonds can be broken and new ones made: the major bonds being due to the equal or unequal sharing of electrons between the nuclei of atoms, with, in extreme cases, the gross inequality of the sharing being sufficient to generate the electrically charged particles we call ions.

All chemical change involves atoms and/or molecules and/or ions colliding in ways which allow the competing strains of the electric force and energy to rearrange these particles' electrons and nuclei into new types of substance, all guided by the inevitable dispersal of energy towards a more even distribution overall.

So chemistry is a frantic dance in which oppositely charged partners are drawn towards one another, while those with like charges are repelled from one another, with the whirling energy of the dance defying these ordering forces as it spreads across the dancefloor.

I savour this rare essence as I think once more of the complex world of chemistry within me and around me—a world of conscious minds and living animals and plants, of rocks and water and wind and fires, and of cars and aeroplanes and plastics and fabrics and a multitude of intricate machines. I savour the essence, and find its simplicity and its power astonishing.

Further reading

Here is a list of some interesting books in which you will find further information about the subjects discussed in this book. Their titles should make the general nature of their contents clear.

Molecules, by Peter W. Atkins, W. H. Freeman, 1987.
The Second Law, by Peter W. Atkins, W. H. Freeman, 1984.
The Forces of Nature, by Paul Davies, Cambridge University Press, 1979.
Chemistry and Chemical Reactivity, by John C. Kotz and Keith F. Purcell, Saunders College Publishing, 1987.
The Problems of Chemistry, by W. Graham Richards, Oxford University Press, 1986.
The Cosmic Code—quantum physics as the language of nature, by Heinz R. Pagels, Penguin, 1982.
The Chemistry of Life, by Stephen Rose, Penguin, 1979.
Introducing Chemistry, by Hazel Rossotti, Penguin, 1975.
The Marvels of the Molecule, by Lionel Salem, VCH (W. Germany), 1987.
The Creation of Life, by Andrew Scott, Basil Blackwell, 1985.
Vital Principles—the molecular mechanisms of life, by Andrew Scott, Basil Blackwell, 1988.
Chemistry in the marketplace, by Ben Selinger, John Murray, 1979

Index

Sapere Aude

TODD WEHR
MEMORIAL
LIBRARY

Viterbo University
900 Viterbo Drive
La Crosse, WI 54601
608-796-3269